Transactions on Computer Systems and Networks

Transactions on Computer Systems and Networks is a unique series that aims to capture advances in evolution of computer hardware and software systems and progress in computer networks. Computing Systems in the present world span from miniature IoT nodes and embedded computing systems to large-scale cloud infrastructures, which necessitates developing systems architecture, storage infrastructure and process management to work at various scales. Present day networking technologies provide pervasive global coverage on a scale and enable a multitude of transformative technologies. The new landscape of computing comprises of self-aware autonomous systems, which are built upon a software-hardware collaborative framework. These systems are designed to execute critical and non-critical tasks involving a variety of processing resources like multi-core CPUs, reconfigurable hardware, GPUs and TPUs which are managed through virtualisation, real-time process management and fault-tolerance. While AI, Machine Learning and Deep Learning tasks are predominantly increasing in the application space the computing system research aim towards efficient means of data processing, memory management, real-time task scheduling, scalable, secured and energy aware computing. The paradigm of computer networks also extends its support to this evolving application scenario through various advanced protocols, architectures and services. This series aims to present leading works on advances in theory, design, behaviour and applications in computing systems and networks. The Series accepts research monographs, introductory and advanced textbooks, professional books, reference works, and select conference proceedings.

L. Yashvanth · Pranav Viswanathan ·
Chandra R. Murthy

Understanding 5G New Radio

An Experimental Approach

L. Yashvanth
Department of Electrical Communication Engineering
Indian Institute of Science Bangalore
Bengaluru, Karnataka, India

Department of Electronic and Electrical Engineering
University College London (UCL)
London, UK

Chandra R. Murthy
Department of Electrical Communication Engineering
Indian Institute of Science Bangalore
Bengaluru, Karnataka, India

Pranav Viswanathan
TETCOS LLP
Bengaluru, Karnataka, India

ISSN 2730-7484 ISSN 2730-7492 (electronic)
Transactions on Computer Systems and Networks
ISBN 978-981-92-0111-2 ISBN 978-981-92-0112-9 (eBook)
https://doi.org/10.1007/978-981-92-0112-9

This Springer imprint is published by the registered company Springer Nature Singapore Pte Ltd.
The registered company address is: 152 Beach Road, #21-01/04 Gateway East, Singapore 189721, Singapore

Preface

The fifth generation (5G) of cellular communication networks marks a significant leap in wireless technology, promising higher data rates, ultra-low latency, and massive device connectivity. As with any technological leap, a deep and intuitive understanding of the underlying principles and system-level trade-offs is essential not only for researchers and engineers but also for students and practitioners working to shape the future of wireless communications.

This book, *Understanding 5G New Radio: An Experimental Approach*, is a product of our experience in academia and industry, and aims to bridge the gap between theory and practice. It offers a unique hands-on approach to understanding 5G New Radio (NR), the global standard for a unified, more capable 5G wireless air interface. Instead of treating 5G purely as a theoretical subject or as a collection of standards documents, we adopt an experimental methodology using the NetSim simulation platform. Each chapter is structured to first build the necessary theoretical foundation and then validate the concepts through carefully designed simulation experiments. This dual-layered approach not only reinforces conceptual clarity but also exposes the reader to the complexities of real-world system behavior. The content spans a wide range of topics fundamental to 5G NR, including MIMO beamforming, pathloss modeling, transport block processing, OFDM numerology, interference and handover modeling, scheduler performance, and massive MIMO. Wherever possible, we emphasize the connection between abstract ideas and their practical implications in system design and performance. To support this hands-on understanding, readers who purchase this book are eligible for a free NetSim license, enabling access to the simulation exercises used throughout the text. An executable version of NetSim v13.3 and the associated experiment configuration files required to execute all the experiments in this book can be found online at https://tetcos.com/5G-Book.html.

This book is intended for graduate students, researchers, educators, and industry professionals interested in 5G NR and wireless systems. While a background in wireless communication principles will aid in comprehension, the book is designed to be accessible to those with varied levels of expertise, particularly through the progressive deepening of concepts.

We hope this book serves not only as a reference but also as a companion in your journey toward mastering 5G systems. As the world begins to explore 6G and beyond, we believe that a strong grasp of 5G NR through both theory and experimentation will serve as a valuable foundation. We welcome feedback, suggestions, and insights from readers, and we look forward to seeing how this book aids in your understanding and innovation.

Bengaluru, India

L. Yashvanth
Pranav Viswanathan
Chandra R. Murthy

Contents

About the Authors

L. Yashvanth received his B.Tech. (Hons.) degree in Electronics and Communication Engineering from the National Institute of Technology, Trichy, India, graduating with a gold medal in 2020, and his Ph.D. from the Indian Institute of Science (IISc), Bengaluru, in 2025. He is currently a research fellow at University College London, UK. His broad research interests include statistical signal processing and its applications to 5G and next-generation wireless communication systems. He is a recipient of the DAAD-WISE Fellowship (2019), the Qualcomm Innovation Fellowship (2021 and 2022), and the Prime Minister's Research Fellowship (PMRF), Government of India (2022). He also received the Best Poster Presentation Award at the PMRF Symposium (2024) and the Prof. Satish Dhawan Research Award (2025) for impactful research in the EECS Division at IISc.

Pranav Viswanathan is the cofounder and CEO of TETCOS LLP (est. 2005) and the chief architect of NetSim, a leading network simulator used by 500+ organizations across 30+ countries. His expertise spans simulation of 5G/6G networks, satellite communications, Mobile Ad Hoc Networks (MANETs), military radios, WiFi, and the Internet of Things (IoT). His current focus areas within NetSim include AI/ML in the RAN, non-terrestrial networking, and MANET military radio networks. Pranav is known for his strong customer-centric approach, translating user requirements into software solutions. With over 20 years of experience in wireless network modeling and simulation, he has traveled globally to support and implement NetSim-based solutions. He earned his B.Tech. from IIT Madras in 2003.

Chandra R. Murthy is a professor in the Department of Electrical Communication Engineering at the Indian Institute of Science (IISc), Bengaluru, India. His research interests include sparse signal recovery, energy harvesting based communication, performance analysis, and optimization of 5G and beyond communication systems. His papers have received Best Paper Awards at NCC 2014 and NCC 2023, and papers coauthored with his students have won Student Best Paper Awards at IEEE ICASSP 2018, IEEE ISIT 2021, and IEEE SPAWC 2022. He currently serves as a senior area editor for the *IEEE Transactions on Information Theory* and has previously served on the editorial boards of several IEEE journals as well as various IEEE technical committees. He is a Fellow of the IEEE and the Indian National Academy of Engineering (INAE).

Chapter 1
Introduction to 5G

1.1 Evolution from 1G to 5G

Wireless cellular communication began in the 1980s with the first generation (1G), which used *analog* modulation to transmit information. Since then, a new "generation" of cellular technology has emerged roughly every decade. The second generation (2G), also known as the Global System for Mobile Communications (GSM), introduced *digital* modulation, enabling text messaging and significantly improving the reliability of voice communications compared to 1G.

The third generation (3G) brought substantial advancements, including higher bandwidths that supported increased data rates, internet access, multimedia storage, and transmission capabilities. For the first time, communication link *bandwidths* expanded into the MHz range. This era also introduced code division multiple access (CDMA), a groundbreaking method that allowed multiple user devices (UEs) to share the same time-frequency resources by assigning unique codes to control inter-user interference (Abu-Rgheff 2007). However, CDMA's scalability was limited by the need for more codes as the number of users grew. Additionally, it required precise real-time power control to prevent interference, and performance issues arose as bandwidths increased, complicating channel estimation and equalization (Li and Stuber 2006). These challenges were addressed in the fourth generation (4G), also known as Long-Term Evolution (LTE), which employed orthogonal frequency division multiplexing (OFDM) to improve system efficiency and performance (Li and Stuber 2006). The latest generation, fifth generation (5G), also called International Mobile Telecommunications for 2020 (IMT-2020), began global deployment in 2020, representing the most advanced stage of wireless cellular technology to date.

L. Yashvanth et al., *Understanding 5G New Radio*, Transactions on Computer Systems and Networks, https://doi.org/10.1007/978-981-92-0112-9_1

1.2 What Does 5G Envision?

The vision for 5G encompasses two primary goals: (1) improving the performance of existing wireless applications and (2) enabling services in new applications to meet the needs of future generations. To achieve these objectives, 5G targets three key use cases:

Enhanced mobile broadband (eMBB) focuses on applications requiring significantly higher data rates than those provided by 4G. A notable feature of eMBB in 5G is the capability to deliver data rates in the range of gigabits per second (Gbps). These rates can be further increased through carrier aggregation, which combines multiple frequency bands to serve a single application simultaneously.

Massive machine-type communications (mMTC) are designed to support a vast number of connected devices, prioritizing the ability to handle millions of devices transmitting short packets over maximizing spectral efficiency. This use case is particularly relevant for Internet of Things (IoT) applications.

Ultra-reliable low-latency communications (URLLC) aim to provide highly reliable, low-latency service for critical applications where minimal delay and extremely low error rates are essential. Examples include remote surgery, autonomous driving, and drone operations.

Further, to evaluate the effectiveness of 5G deployments, the International Telecommunication Union (ITU) has defined eight key performance capabilities, ensuring each use case meets its specific requirements effectively. These capabilities and the use cases for which each metric is relevant are pictorially captured in Fig. 1.1 (IMT Vision 2015). A key point to note is that 5G is not envisioned to meet the requirements of all three use cases simultaneously.

1.3 Towards 5G Deployment and Its Features

The first official technical specifications for global 5G system deployment were released in 2018 as part of "*5G New Radio (NR), Release 15*" (ETSI 2023). This release comprises a series of standards documents outlining the specifications that 5G systems must meet. Below, we discuss some of the key features of the 5G standard.

1.3.1 Components of 5G NR

The 5G network consists of three main components: user equipment (UE), the radio access network (RAN), and the 5G core network (CN).

- *User Equipment (UE)*: This refers to the devices or handsets used by end users (e.g., individuals) to access wireless services.

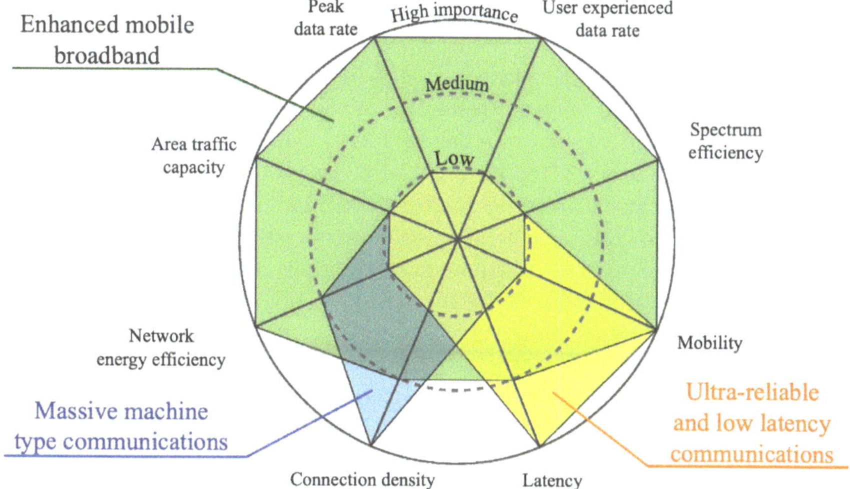

Fig. 1.1 Primary use cases of 5G and the corresponding capability requirements. Here, each vertex represents a requirement, and each color represents a use case. The requirements to be met by a particular use case are marked by coloring the vertex corresponding to the use case. Courtesy IMT Vision (2015)

- *Radio Access Network (RAN)*: The RAN encompasses all functionalities related to the UE's radio connectivity with the 5G access points (APs). These APs, referred to as gNBs (base stations), are deployed by mobile service providers to establish and maintain radio connections.
- *5G Core Network (CN)*: The core network manages essential services for users through various network functions, including authentication, session management, and control and user plane operations. Additionally, the CN handles tasks such as load balancing across gNBs and managing seamless handovers as users move within the network's coverage area (Penttinen 2019).

1.3.2 Spectrum for 5G

To meet the demand for high data rates, the 5G standards define two distinct frequency ranges (FRs) for operation based on use cases and deployment scenarios: FR-1 and FR-2. The FR-1 band, also known as the sub-6 GHz band, includes lower frequency ranges (410 MHz–7.125 GHz), while the FR-2 band supports transmissions in the millimeter wave (mmWave) spectrum, spanning frequencies from 24.25 to 52.6 GHz (5G NR 2017). The FR-2 bands were newly introduced with the 5G specifications. The FR-1 band is ideal for wide coverage due to its superior propagation characteristics, supporting channel bandwidths of up to 100 MHz on a single carrier (with carrier aggregation, it could be higher). In contrast, the FR-2 band is designed for high data rates in hotspot areas, utilizing wider bandwidths of up to 400 MHz without carrier aggregation. However, the FR-2 band presents unique challenges, such as higher attenuation losses and directional propagation, necessitating advanced engineering solutions to develop robust and efficient systems for operation in these bands (Huang and Wang 2011). Another notable feature of the 5G architecture is the ability to implement variable OFDM subcarrier bandwidths by dynamically selecting different numerologies (Zaidi et al. 2016; Yang et al. 2021a). The introduction of variable numerologies accommodates diverse deployment scenarios, ranging from large cells in rural areas to microcells in heterogeneous networks with high data rate demands and low-latency communication requirements. Variable numerologies specifically allow for different cyclic prefix durations across the FR-1 and FR-2 bands. Additionally, since changing the subcarrier bandwidth alters the OFDM slot duration, variable numerologies enable support for application-specific packet lengths without compromising the performance of other applications. However, this flexibility introduces engineering trade-offs, such as balancing latency and throughput performance for delay-sensitive applications.

1.3.3 Enhanced Logical Multiplexing

An exciting feature of the 5G specifications is network slicing (Kazmi et al. 2019). This capability allows network operators to create multiple independent logical channels on the same network infrastructure, enabling various applications and services with diverse quality of service (QoS) requirements to share the same physical resources. For example, this technology facilitates the efficient coexistence of eMBB and URLLC use cases on a single physical infrastructure and share common resources while ensuring that each meets its respective QoS requirements.

1.3.4 Coexistence with Previous Technologies

The real-time deployment of 5G systems must be scalable and coexist with previous generations, such as 3G and 4G, especially when deployed in the same or overlapping frequency bands. To achieve this, mechanisms are integrated into 5G to ensure compatibility with older wireless standards. For example, 5G supports dual connectivity, allowing a single UE to connect simultaneously to two APs—one operating on the 4G stack for the control plane traffic and the other on the 5G stack for the data plane traffic. This approach guarantees reliable connectivity via 4G links while maximizing the benefits of 5G services (Nath et al. 2023).

1.3.5 Channel Measurement Methods

A notable addition to the 5G standards is the introduction of beam management, a procedure designed to establish link connectivity in the FR-2 bands. Due to the significant propagation losses experienced in mmWave frequencies compared to sub-6 GHz bands, deploying a massive number of antennas at the gNBs is recommended. These antennas form directional beams to maintain connectivity, enabling the implementation of massive multiple-input multiple-output (MIMO) techniques (Ngo 2015). However, this requires a deep understanding of the limitations and benefits of massive MIMO systems. Additionally, due to the highly directional nature of the channels, multiple beams are needed to cover a geographic area, which can lead to delays in establishing a connection between the gNB and the UE. To address this, 5G incorporates hierarchical beam management protocols designed to identify the correct beam as quickly as possible using the physical broadcast channel (PBCH) resources (Yang et al. 2021b).

1.3.6 Robust Transmission Techniques

Another key difference from 4G systems is that the 5G transport layer employs LDPC codes and polar codes for forward error correction in the data and control channels, respectively (Xu and Yuan 2022). As a result, these codes are expected to deliver near-optimal capacity-achieving solutions compared to their 4G counterparts, significantly boosting data rates.

While there are many other differences in the operation of an end-to-end 5G system compared to its predecessor generations, we have highlighted only some of the key and fundamental features of the 5G standards above. Readers are suggested to refer (Boldi et al. 2018; Dahlman et al. 2020; Lei et al. 2021; Ahmadi 2019), and obtain detailed explanations of various other theoretical aspects of 5G NR.

1.4 Importance of Understanding 5G

The industrial research community has begun exploring the potential use cases that 6G is expected to address (IMT-2030), with global 6G deployment anticipated to begin around 2030 (International Telecommunication Union 2023). New use cases such as integrated sensing and communications (Masouros et al. 2023), merging of artificial intelligence to wireless applications (Goldsmith et al. 2022), and immersive communication have been proposed in addition to enhanced versions of the three primary use cases of 5G, making a total of six primary use cases for the 6G. Further, 6G is also conceived to provide seamless service to these use cases through 14 different capabilities, of which eight are derived from 5G.

However, much research is required to develop protocols and standards for meeting the new and enhanced requirements of IMT-2030. Several emerging technologies are being considered and are still in the early research phase. Some of these include cell-free communications (Demir et al. 2021), intelligent reflecting and holographic surfaces (Wu et al. 2024), full-duplex technology (Song et al. 2017), non-terrestrial communications (Heyn et al. 2022), extremely large (XL)-MIMO, and so on (Jiang and Luo 2022). Extensive field measurements need to be conducted to assess the effectiveness of these technologies in real-world scenarios. Following this, feasibility studies will be carried out to integrate these technologies into the 6G standards, evaluating their performance and trade-offs. Therefore, to fully grasp the key advancements that 6G would introduce and how it will surpass its predecessors, it is essential to have a thorough understanding of 5G technologies.

This book aims to serve as an introduction to 5G, taking an experimental approach through a series of simulation experiments that align with the 3GPP specifications for 5G NR.

References

5G NR (2017) UE radio transmission and reception; part 1: range 1 standalone, v 15.2.0 rel 15. 3rd Generation Partnership Project (3GPP), France

Abu-Rgheff MA (2007) Introduction to CDMA wireless communications. Academic Press

Ahmadi S (2019) 5G NR: architecture, technology, implementation, and operation of 3GPP new radio standards. Elsevier Science

Boldi M, Queseth O, Marsch P, Bulakci Ö (2018) 5G system design: architectural and functional considerations and long term research. Wiley

Dahlman E, Parkvall S, Skold J (2020) 5G NR: the next generation wireless access technology. Elsevier Science

Demir ÖT, Björnson E, Sanguinetti L (2021) Foundations of user-centric cell-free massive MIMO. Now Publishers

ETSI (2023) NR; NR and NG-RAN overall description; stage 2 (3GPP TS 38.300 version 15.15.0 release 15). ETSI TS 138 300 V15.15.0, July 2023. [Online]. Available: https://www.etsi.org/deliver/etsi_ts/138300_138399/138300/15.15.00_60/ts_138300v151500p.pdf

Goldsmith A, Gündüz D, Poor HV, Eldar YC (2022) Machine learning and wireless communications. Cambridge University Press

Heyn T, Hofmann A, Raghunandan S, Raschkowski L (2022) Non-terrestrial networks in 6G. In: Shaping future 6G networks: needs, impacts, and technologies. IEEE-Wiley Press, pp 101–116

Huang K-C, Wang Z (2011) Millimeter wave communication systems. Wiley

IMT Vision (2015) Framework and overall objectives of the future development of IMT for 2020 and beyond. Recommendation ITU-R M.2083-0. Geneva

International Telecommunication Union (2023) Framework and overall objectives of the future development of IMT for 2030 and beyond. International Telecommunication Union, Radiocommunication Sector, Geneva

Jiang W, Luo F-L (2022) 6G key technologies: a comprehensive guide. Wiley

Kazmi S, Khan L, Tran N, Hong C (2019) Network slicing for 5G and beyond networks, vol 1. Springer, Berlin

Lei W et al (2021) 5G system design: an end to end perspective. Springer International Publishing

Li YG, Stuber GL (2006) Orthogonal frequency division multiplexing for wireless communications. Springer Science & Business Media

Masouros C, Liu F, Eldar YC (2023) Integrated sensing and communications. Springer Nature, Singapore
Nath SV, Simlai A, Kara O (2023) Mastering 5G network design, implementation, and operations. Packt Publishing
Ngo HQ (2015) Massive MIMO: fundamentals and system designs, vol 1642. Linköping University Electronic Press
Penttinen JT (2019) 5G explained: security and deployment of advanced mobile communications. Wiley
Song L, Wichman R, Li Y, Han Z (2017) Full-duplex communications and networks. Cambridge University Press
Wu Q, Duong TQ, Ng DWK, Schober R, Zhang R (2024) Intelligent surfaces empowered 6G wireless network. IEEE-Wiley Press
Xu J, Yuan Y (2022) Channel coding in 5G new radio. CRC Press
Yang B, Zhang X, Zhang L, Farhang A, Xiao P, Imran MA (2021a) Windowed OFDM for mixed-numerology 5G and beyond systems. In: Radio access network slicing and virtualization for 5G vertical industries. IEEE Wiley Press, pp 43–61
Yang P, Wu W, Zhang N, Shen X (2021b) Millimeter-wave networks: beamforming design and performance analysis. Springer International Publishing
Zaidi AA et al (2016) Waveform and numerology to support 5G services and requirements. IEEE Commun Mag 54(11):90–98

Chapter 2
Learning 5G Using NetSim

This book presents a unique approach to understanding 5G systems via a series of simulation-based experiments conducted using a network simulation tool called NetSim. However, before delving into the details of how NetSim can be used to simulate and study 5G systems, we first discuss the general principles of network simulation and its essential characteristics.

2.1 Simulation

A simulation is, roughly speaking, a software-based abstraction, or a digital twin, that imitates the operation of a system over time. Simulation involves the generation of an artificial history of the temporal evolution of a system and the observation of that artificial history to draw inferences. The behavior of a system as it evolves over time is studied by developing a simulation model. Once developed and validated, a model can be used to investigate a wide variety of "what if" questions about the real-world system quickly and cost-effectively. Potential changes to the system can be simulated to predict their impact on system performance. Simulation can be used to study systems in the design stage before such systems are built. Thus, simulation modeling can be used both as an analysis tool for predicting the effect of changes to existing systems and as a design tool to predict the performance of new systems under varying sets of circumstances, provided the simulation captures the essential characteristics of the real system (Banks et al. 2010).

L. Yashvanth et al., *Understanding 5G New Radio*, Transactions on Computer Systems and Networks, https://doi.org/10.1007/978-981-92-0112-9_2

2.2 Discrete Event Simulation

A discrete event simulation attempts to predict or reproduce the behavior of the system of interest by decomposing the behavior into a sequence of state changes. Each state change for the system, and each instant in time at which a state change might occur, is referred to as an event. Events are tagged with simulation time and stored in a structure called the event list. The simulator executes these events in the order of monotonically increasing simulation time. After each event's completion, the simulator advances to the next event or terminates if no more events remain in the queue.

2.2.1 Stochastic Modeling

Randomness, or stochasticity, is essential in capturing the behavior of real-world systems. In a real-world system, the wireless channels, data patterns, etc., are necessarily random, and a realistic evaluation of their behavior requires stochastic modeling. Stochastic models are based on random trials, in contrast to deterministic models that always produce the same output for a given starting condition.

A stochastic model recognizes and incorporates the inherently random nature of certain system variables. Discrete event simulations rely on stochastic modeling of system properties to capture behavior that is not deterministic but can only be characterized statistically. Stochastically modeled attributes depend on a random number source to select outcomes that drive their behavior. The outcome of simulations, therefore, depends on the sequence of numbers drawn from a random number generator. The random number sequence is, in turn, governed by the seed value of the random number generator. Therefore, one usually cannot rely on a single discrete event simulation experiment to reach conclusions about the simulated system. Each individual simulation represents only one possible outcome of the modeled system. Some simulations may produce results that are outliers.

Proper design of a simulation study generally requires that one execute a series of simulations with varying seeds to obtain a set of results, which, together, can be used to characterize the system behavior and provide reliable answers to questions of interest. The set of results is generally used to find a range, a mean, or a standard deviation, and to compute confidence intervals. Regrettably, too many simulation studies ignore this important principle and base their answers on just one simulation result or a small number of simulation runs. This can lead to flawed conclusions.

2.2.2 *Confidence in Simulation Results*

As explained above, NetSim's models incorporate the stochastic behavior of real-world networks. Results are dependent on the initial seeding of the random number generator, and because a particular random seed selection can potentially result in an anomalous or non-representative behavior, it is important for each model configuration to be exercised with several random number seeds to determine the standard or typical behavior as well as the variations around the typical behavior.

The field of statistics provides methods for calculating confidence in an estimate, based on a trial or series of random trials. To calculate confidence intervals, one can do the following:

1. Run N simulations for the same model configuration, with a different initial seed (for the random number generator) for each run.
2. For any output metric X, calculate the sample mean (average) $\overline{X}$ of the N output values.

$$\overline{X} = \frac{1}{N}(X_1 + X_2 + \cdots + X_N).$$

3. Calculate the standard deviation σ of the N samples.

$$\sigma = \sqrt{\left(\frac{1}{N-1}\right) \times \sum_{i=1}^{N} (X_i - \overline{X})^2}.$$

4. Confidence interval limits can be expressed as

$$\Theta_{\text{lower}} = \overline{X} - z_\alpha \times \left(\frac{\sigma}{\sqrt{N}}\right), \quad \Theta_{\text{upper}} = \overline{X} + z_\alpha \times \left(\frac{\sigma}{\sqrt{N}}\right).$$

This statement can be thought of as assigning a probability to the condition that the true mean, μ, is within a particular interval around the sample mean, $\overline{X}$, as shown below.

$$\text{Prob}\left[\overline{X} - z_\alpha \times \left(\frac{\sigma}{\sqrt{N}}\right) < \mu < \overline{X} + z_\alpha \times \left(\frac{\sigma}{\sqrt{N}}\right)\right] = \alpha.$$

In Table 2.1, we provide the value of the confidence interval parameter z_α for different confidence levels or probabilities α. The above expression for confidence assumes the distribution of the output metric is Gaussian (also called a Normal distribution), which is true (from the celebrated Central Limit Theorem) if the number of runs $N \geq 30$. If the number of repetitions is less than 30, then it is better to use the t-statistic based confidence interval (details can be found in standard textbooks on statistics).

Table 2.1 Value of z_α for different confidence intervals

Confidence level (α) (%)	z_α
99	2.575
98	2.327
95	1.960
90	1.645
80	1.282

2.2.3 Simulation Time and its Relation to Real Time (Wall Clock)

Earlier, we discussed how events in a simulation are arranged and executed in sequential order of simulation time. But what exactly do we mean by simulation time? It is important to understand that the notion of time in a simulation is not directly related to the actual time that it takes to run a simulation (as measured by a wall clock or the computer's own clock), but is a variable maintained by the simulation program. NetSim uses a virtual clock which ticks virtual time. Virtual time starts from zero and progresses as a positive real number.

In NetSim, (this virtual) time is a global parameter. All components of the network share the same time throughout the simulation independently of where they are physically located or how they are logically connected to the network. Put differently, there are no differences in the local time of the communication network components.

This virtual time is referred to as simulation time to clearly distinguish it from real (wall clock) time. NetSim is a discrete event simulator (DES), and in any DES, the progression of the model over simulation time is decomposed into individual events where changes can take place. The flow of time is only between events and is not continuous. Simulation time does not progress during an event, but only between events. In fact, the simulation time is always equal to the time at which the current event occurs. Thus, simulation time can be viewed as a variable that "jumps" to track the time specified for each new event.

For example, if we wish to measure the throughput at a user equipment connected to a 5G network, we run the simulation for a certain duration of simulation time, and count how many bits of data were delivered. Then, we divide the number of bits by the simulation time to obtain an estimate of the throughput. For instance, if we consider 10 s of simulation time, depending on the size of the network and how fast the computer is, it may require a different amount of time as per the wall clock to run a 10-s simulation, but we would still divide the number of bits delivered by 10 to obtain the throughput. In this way, NetSim computes the throughput independent of the speed with which the simulation runs.

2.3 NetSim: A Discrete Event, End-to-End, System Level, 5G Simulator

NetSim (www.tetcos.com) is a platform for simulating 5G networks and analyzing the network's behavior and performance. The NetSim workflow includes:

- Network design.
- Simulation execution.
- Output data collection and analysis.

For advanced users, there are options to:

- Develop custom algorithms.
- Integrate with external software.

NetSim finds typical applications in studying network capacity and scalability, evaluating network performance, and generating synthetic data for machine learning applications without the need for costly physical equipment.

Typical users of NetSim include:

- Government sectors, including military, defense, space, and telecommunications regulators.
- Enterprises, such as those in telecommunications (particularly 5G), critical infrastructure, autonomous vehicles, and equipment manufacturing.
- Educational institutions for teaching, lab experimentation, and research and development efforts.

2.4 Challenges in Building a 5G Simulator

Ensuring accuracy and realism in modeling a complex and dynamic system comes with several challenges:

- Scale
  - Tens of gNBs and 100s of UEs.
  - Multi Gbps data speeds at each gNB.
  - Need to compute pathloss and interference between all pairs of gNBs and UEs to accurately model the wireless aspect of cellular communications.
- Granularity
  - Scheduling data during every transmit time interval (TTI), which could be as low as 0.125 ms.
- Complexity
  - Interlinked stochastic computations for pathloss, Shadowing, Fast fading MIMO, Beamforming, Mobility, Handover, Scheduling, etc.

- Computation of Logs
  - Measurements: every slot.
  - Packet and event trace: every packet.
- Testing and Verification
  - Extensive analytical studies.
  - Curse of dimensionality: Hundreds of input parameters, denoted by p, and a wide range of possible values, denoted by ν, for each input. Then, $\mathcal{O}(\nu^p)$ is the number of test scenarios that need to be supported.
  - Reproducibility and backward compatibility with every new release.

2.5 5G Physical Layer (PHY) Implementation in NetSim

NetSim is a packet-level simulator for simulating the performance of end-to-end applications over various packet transport technologies. NetSim can scale to simulating networks with 100s of end-systems, routers, switches, etc. It provides estimates of the statistics of application-level performance metrics such as throughput, delay, packet loss, and statistics of network-level processes such as buffer occupancy, collision probabilities.

Of all the wireless access technologies implemented in NetSim, the most sophisticated is 5G NR, in which the physical layer utilizes a variety of techniques that go well beyond 4G LTE. To achieve a scalable simulation that can execute in a reasonable time on desktop computers, the details of the physical layer techniques are abstracted out. These include multiple subcarrier bandwidths in the same system, slot lengths that depend on the subcarrier bandwidth, flexible time-division duplexing, a wide range of constellation sizes and coding rates, multiuser MIMO-OFDM, etc. Particularly, NetSim takes the design approach of replacing these with symbol-level models, where only the statistics of the effective stochastic channel gains, and the statistics of the effective noise and interference are modeled in the simulation. Such models then permit calculating the required bit error and code block error rates at speed.

Reference

Banks J, Carson J, Nelson B, Nicol D (2010) Discrete-event system simulation. Pearson

Chapter 3
Overview of This Book

Each chapter of this book explores a specific aspect of 5G NR and demonstrates it in detail via a series of simulation experiments performed on NetSim v13.3. A summary of the chapters and their contents is provided below.

A. Chapter 4: MIMO Beamforming in 5G Communications: A Start to Multi-antenna Systems.

This chapter demonstrates the role of multiple antenna technologies in 5G NR. It examines how the beamforming gain and throughput vary with the number of antennas deployed at various nodes in the system. The performance is analyzed for multiple-input single-output (MISO) and single-input multiple-output (SIMO) scenarios, accounting for practical aspects such as channel quality index (CQI) feedback and modulation and coding scheme (MCS) index selection. The chapter highlights that beamforming gain increases linearly while throughput grows log-linearly with the number of antennas.

B. Chapter 5: Understanding 5G NR Pathloss Models.

This chapter provides a foundational understanding of various pathloss models outlined in the 3GPP 5G NR standards. The pathloss is influenced by the physical characteristics of the propagation environment and varies across different locations. These parameters are crucial for designers to assess link budget requirements and plan cell deployments effectively. The chapter offers a detailed discussion of how the pathloss depends on system-level parameters such as the gNB-UE distance, UE/gNB heights, and operating frequency. It concludes with a practical exercise on solving a cell-planning problem for rural areas based on specified link budget parameters.

C. Chapter 6: Understanding 5G NR PHY: Transport Block Processing.

This chapter offers a fundamental understanding of the end-to-end transport block processing mechanism in the 5G NR physical layer (PHY) within the OFDM framework. It begins with an overview of typical resource block allocation strategies on the

L. Yashvanth et al., *Understanding 5G New Radio*, Transactions on Computer Systems and Networks, https://doi.org/10.1007/978-981-92-0112-9_3

time–frequency OFDM grid. A detailed, step-by-step explanation of physical layer processing follows, illustrating how a transport block packet is processed through the complete transmit-receive chain. The chapter concludes with an in-depth discussion on throughput evaluation in a 5G system, supported by a numerical example, and highlights how the throughput depends on factors such as transmit power, MIMO layer count, and system bandwidth.

D. Chapter 7: Performance of 5G Single-user MIMO Orthogonal Frequency Division Multiple Access (SU-MIMO-OFDMA).

This chapter examines and evaluates the performance of single-user (SU) MIMO systems within an OFDMA-based 5G NR framework. Specifically, it analyzes SU-MIMO performance in two scenarios: (1) a system where a single UE is allocated all frequency resources and MIMO layers and (2) a system with multiple UEs, where each UE is assigned all MIMO layers, but the UEs multiplexed across different subcarriers of the OFDM system. A comparative analysis of these scenarios is conducted, which provides insights into the overall functioning of a MIMO-OFDMA system.

E. Chapter 8: Impact of Numerology on End-to-End Latency and Throughput in 5G NR OFDM systems.

This chapter explores the impact of variable subcarrier bandwidth (spacing) in OFDM, achieved by altering the numerology in 5G NR. Changes in subcarrier spacing affect the time duration of OFDM symbols, which in turn influences the overall performance. Specifically, the chapter examines how numerology impacts the end-to-end latency and achievable throughput under two transport layer protocols: (1) transmission control protocol (TCP) and (2) user datagram protocol (UDP). A comprehensive performance evaluation is presented, supported by relevant simulation results and detailed justifications.

F. Chapter 9: Understanding the Role of Interference in 5G Network Design: Interference Modeling in Handover Procedures.

This chapter examines the role of interference in 5G NR and its impact on system performance. Typically, a UE experiences interference from adjacent cell BSs, rendering the system interference-limited at the cell-edge. This phenomenon is illustrated by analyzing the handover process, where a mobile UE transitions from one BS to another as it crosses cell boundaries. In particular, the chapter highlights how interference influences the handover point (the location where handover occurs) of a UE.

G. Chapter 10: Understanding Handover Mechanisms in 5G NR.

This chapter provides an in-depth analysis of system-level mechanisms and performance variations during the handover procedure studied in the previous chapter. Specifically, the following aspects are considered: (1) the complete packet trace during the handover signaling phase and (2) the variation in throughput and end-to-end latency during the handover phase. The influence of system and handover design parameters on these performance metrics is illustrated. The chapter concludes with interesting exercises that study the impact of the so-called handover margin on the behavior of the system.

H. Chapter 11: On the Impact of MAC Schedulers for 5G NR: Design and Analysis.

This chapter explores various user scheduling algorithms as specified in 5G NR, which play a pivotal role in optimizing system performance. The focus is on three widely used scheduling algorithms: (1) max-rate, (2) proportional-fair, and (3) round-robin. For each algorithm, the design principles and trade-offs are analyzed based on multiple performance metrics. A performance evaluation is presented under varying wireless channel conditions. The chapter also concludes with exercises that highlight the importance of opportunistic communications and the associated multi-user diversity benefits.

I. Chapter 12: Performance of Large MIMO Systems in 5G NR: Theory and Experiments.

This final chapter delves into the theoretical performance limits of large MIMO systems, commonly referred to as Massive MIMO, which employ a substantial number of antennas at the gNBs. Leveraging concepts from random matrix theory, the chapter analyzes the eigenvalue distribution of large MIMO channels, a critical determinant of overall system throughput. The study focuses on understanding channel hardening—a phenomenon where the randomness of the wireless channel diminishes due to the use of numerous antennas. Key aspects, including the asymptotic convergence of the condition number of the channel matrix and the alignment of the eigenvalue spectrum with the well-known Marchenko-Pastur distribution, are examined. These findings offer valuable insights into the behavior and design of large MIMO systems in 5G networks.

Chapter 4
MIMO Beamforming in 5G Communications: A Start to Multi-antenna Systems

4.1 Objective

Consider a 5G communication between a gNB and a single UE over a fading channel. Setup the simplest multiple-antenna cases, namely: (1) multiple-input single-output (MISO) and (2) single-input multiple-output (SIMO), and investigate the questions:

- How does beamforming gain vary with antenna count?
- How does throughput vary with antenna count?

4.2 Theory

Multiple-input multiple-output (MIMO) is a method for increasing the capacity of the wireless channel using multiple transmitting and receiving antennas. Multiple antennas exploit the spatial dimension, i.e., multiple paths from transmitter to receiver, under suitable spacing within the antenna array on each side and channel scattering conditions.

Consider a $N_r \times N_t$ MIMO system where N_t is the number of transmit antennas and N_r is the number of receive antennas. The simplest MIMO instantiations are when:

- $N_t = 1$, a special case where the MIMO system simplifies into a single-input multiple-output (SIMO) channel, and
- Reciprocally, $N_r = 1$, a special case where the MIMO system reduces to a multiple-input single-output (MISO) channel.

In both SIMO and MISO, the number of layers (i.e., spatial streams with independent data) is given by $\min(N_r, N_t)$, which equals 1.

SIMO: SIMO occurs when the transmitter has a single antenna, and the receiver has multiple antennas. The signal received on multiple antennas is combined to maximize

L. Yashvanth et al., *Understanding 5G New Radio*, Transactions on Computer Systems and Networks, https://doi.org/10.1007/978-981-92-0112-9_4

an appropriate metric. For example, when the goal is to maximize the received SNR, under additive white Gaussian noise, the optimal receiver is called the maximal ratio combining. In the case of fading channels (e.g., with Rayleigh fading), the channels between the transmitter and the different receive antennas are modeled as independent and identically distributed with unit variance entries; this is shown in Fig. 4.1, where h_{i1} represents the channel coefficient between the transmitting antenna and the ith receiving antenna. In this case, maximal ratio combining uses the complex conjugate of the channel coefficients as the weights to combine the signals, and in turn, this provides *receive diversity* gain. The *average* SNR at the receiver improves by $10 \log_{10}(N_r)$. However, the improvement in SNR is not exactly the same for *every* channel instantiation, since the channel is random. In the experiment, we will quantify the improvement in the data rate as we vary N_r.

MISO: The gain and phase of the signal from each of the transmit antennas is adjusted so that they add constructively at the receiver, yielding an N_t fold power gain on average. As with SIMO, the instantaneous SNR gain, however, will differ from N_t since the channel is random. So, the question is, given a choice between having multiple antennas at either the transmitter or the receiver, which option yields better improvement in the throughput? Or will they be the same? Note that, with a single-antenna receiver, only one data stream is transmitted, and therefore multiple antennas offer only a diversity gain, but not a multiplexing gain. Now, practically speaking, is it better to have multiple antennas at the base station or at the user? In terms of antenna placement, there is more space to install antennas at the base

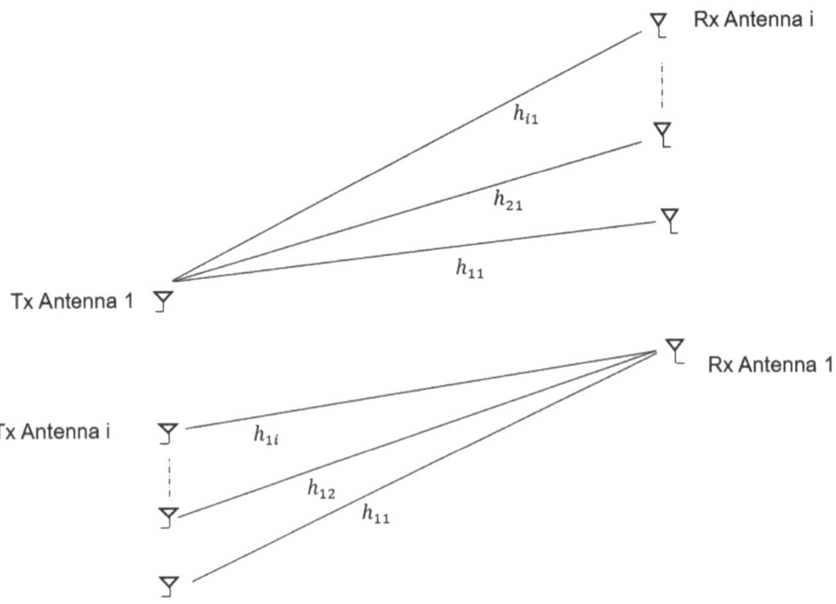

Fig. 4.1 SIMO and MISO architectures

station. In addition, the signal processing capability of the base station is much higher than that of a mobile phone. Thus, multiple antennas at the base station are easier to implement than multiple antennas at the mobile phone.

The Rayleigh Fading Channel: For a transmitter (gNB) with N_t antennas and a receiver with N_r antennas, the $N_r \times N_t$ baseband channel gain matrix (to model fading between every transmit-receive antenna pair) has complex Gaussian distributed elements (Tse and Viswanath 2005). The standard model (under the assumption of Rayleigh fading) is that the complex elements are statistically independent across antennas, and each element is a circularly symmetric complex Gaussian distributed with zero mean and unit variance. We denote this matrix by $\boldsymbol{H}$.[1]

For the channel matrix $\boldsymbol{H}$ defined above, consider the complex Wishart matrix, defined as:

$$\boldsymbol{W} = \boldsymbol{H}\boldsymbol{H}^{\dagger}, \quad N_r < N_t,$$

$$\boldsymbol{W} = \boldsymbol{H}^{\dagger}\boldsymbol{H}, \quad N_r \geq N_t,$$

where $(\cdot)^{\dagger}$ stands for the Hermitian, or conjugate transpose, operation. Therefore, letting $m = \min(N_r, N_t)$, $\boldsymbol{W}$ is an $m \times m$ nonnegative definite matrix with eigenvalues $\lambda_1 \geq \lambda_2 \geq \lambda_3 \geq \cdots \geq \lambda_L > 0 = \lambda_{L+1} = \cdots = \lambda_m$. It is these eigenvalues that determine the gains in the parallel SISO models that arise from eigen-beamforming at the transmitter and receiver (Goldsmith 2005).

NetSim permits the user to enable or disable a stochastic fading model. Fading is modeled by the elements of H varying over time, taking on a new instantiation after each *coherence time*. Such time variation results in the eigenvalues of W also being time-varying. NetSim models such time variation by letting the user define a coherence time during which the eigenvalues are kept fixed. For each (N_r, N_t) value, NetSim maintains a list of samples of eigenvalues for the corresponding Wishart matrix.

4.3 Network Scenario

As an illustration, we provide, in Fig. 4.2, the network topology of the considered MIMO system.[2]

---

[1] Note that H is a *random* matrix. It is the distributions and resulting expectations that determine the average performance.

[2] For e.g., NetSim UI would display the network topology shown in the figure when you open the example configuration file.

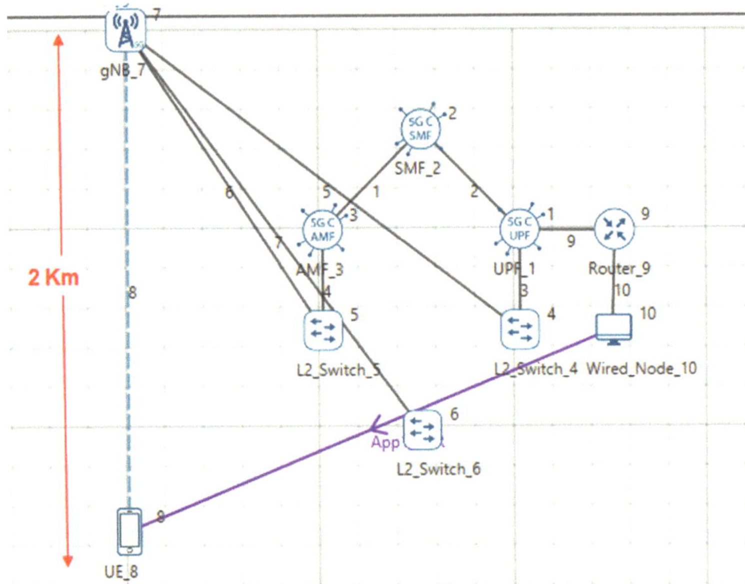

Fig. 4.2 Network topology in this experiment

4.4 Network Configuration

The following parameters were configured in the network setup:

1. The gNB was set with the properties, as shown in Table 4.1.
2. The UE was set with the properties, as shown in Table 4.2.
3. See the Appendix for details on the experimental implementation in NetSim v13.3.

4.5 Sample Results

MISO: Varying Tx Antenna Count in the gNB and 1 Rx Antenna in the UE

See Table 4.3.

SIMO: Varying Rx Antenna Count in the UE and 1 Tx Antenna in the gNB

See Table 4.4.

Beamforming Gain Plot

See Fig. 4.3.

Table 4.1 gNB properties

gNB—interface 5G_RAN parameters	
gNB height	10 m
Tx power	40 dBm
Duplex mode	TDD
CA type	SINGLE BAND
CA configuration	n78
DL: UL ratio	4:1
Numerology	0
Channel bandwidth (MHz)	10
Tx antenna count	Varied from 1 to 128
Rx antenna count	1
MCS table	QAM64
CQI table	TABLE1
Pathloss model	3GPPTR38.901-7.4.1
Outdoor scenario	Urban macro
LOS NLOS selection	User defined
LOS probability	0 (NLOS)
Shadow fading model	None
Fading and beam forming	RAYLEIGH with EIGEN beamforming
Coherence time (ms)	10
Additional loss model	None

Table 4.2 UE properties

UE interface 5G RAN	
Tx power	23 dBm
UE height	1.5 m
Tx antenna count	1
Rx antenna count	Varied from 1 to 16

4.6 Discussion

From the tabulated results given in Tables 4.3 and 4.4, we observe that

- An increase in beamforming gain, as
  - N_t increases, in the MISO case, and as
  - N_r increases, in the SIMO case.
- The beamforming gain when N_t varies (with $N_r = 1$) is precisely the same as when N_r varies (with $N_t = 1$).

Table 4.3 NetSim simulation output showing the throughput, average beamforming gain, and the upper bound (from Jensen's inequality) on the beamforming gain for an $1 \times N_t$ MISO channel

gNB_Tx antenna count	UE_Rx antenna count	Throughput (Mbps)	Average beamforming gain (dB). # layers = 1	Upper bound on the beamforming gain (dB)	Pathloss (dB)	Average SNR (dB)	Average CQI Index	Average MCS Index
1	1	0.14	− 2.27	0	153.54	− 12.00	0.63	0.13
2	1	0.66	1.70	3.01	153.54	− 8.01	1.49	0.61
4	1	1.82	5.41	6.02	153.54	− 4.30	2.89	2.02
8	1	3.60	8.66	9.03	153.54	− 1.06	4.40	4.82
16	1	6.41	11.91	12.04	153.54	2.19	6.28	9.00
32	1	10.01	14.98	15.05	153.54	5.26	7.94	12.89
64	1	14.49	18.04	18.06	153.54	8.32	9.95	17.84
128	1	18.92	21.05	21.07	153.54	11.33	11.39	20.78

Table 4.4 NetSim simulation output showing the throughput, average beamforming gain, and the upper bound (from Jensen's inequality) on the beamforming gain for a $N_r \times 1$ SIMO channel

gNB_Tx antenna count	UE_Rx antenna count	Throughput (Mbps)	Average beamforming gain (dB)	Upper bound on the beamforming gain (dB)	Pathloss (dB)	Average SNR (dB)	Average CQI Index	Average MCS Index
1	1	0.14	− 2.27	0	153.54	− 12.00	0.63	0.13
1	2	0.66	1.70	3.01	153.54	− 8.01	1.49	0.61
1	4	1.82	5.41	6.02	153.54	− 4.30	2.89	2.02
1	8	3.60	8.66	9.03	153.54	− 1.06	4.40	4.82
1	16	6.41	11.91	12.04	153.54	2.19	6.28	9.00

N_r is limited to 16 since this is the maximum antenna count supported in UEs in NetSim

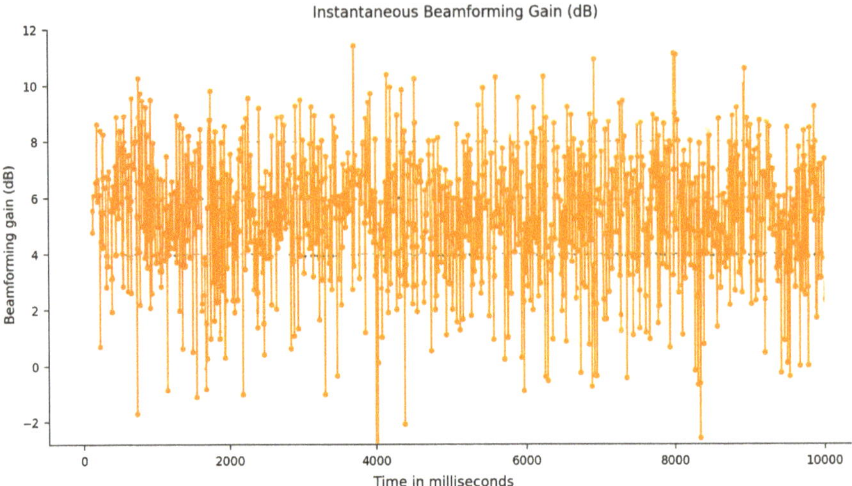

Fig. 4.3 NetSim simulation output showing the variation in the beamforming gain (dB) over the course of the simulation for 4×1 system. The beamforming gain changes after every coherence time

Next, we turn to the question a network engineer would be interested in: how does beamforming impact throughput? While link-level simulators may perform the beamforming computations and provide the SNR at a link level, the power of a "system" level simulator like NetSim lies in its ability to compute the impact of link-level factors (such as beamforming) on the system (or network).

Since the distance between the gNB and UE is fixed, the common pathloss for all Tx–Rx antenna pairs is the same. This common pathloss is factored out (or "pulled out") from the individual $N_t - N_r$ pathloss calculations. With this factorization done, the only parameter affecting SNR is the *channel fading*, the effect of which shows up in the output as *beamforming gain*. Quite simply, as the (average) beamforming gain increases, the (average) SNR proportionally increases. Notice that every time the *antenna count is doubled*, the *SNR increases by* $\approx$ *3 dB*, i.e., a factor of 2 (which matches intuition). An increase in SNR improves the channel quality (the CQI), and thereby, a higher modulation and coding scheme (MCS) is chosen for data transmission, leading to higher throughputs.

Remark. Note that these are "average" arguments: in practice, since the fading coefficients are random, one does not obtain a 3 dB improvement by doubling the number of antennas for every channel instantiation, which can be clearly seen from Fig. 4.3. Consequently, the improvement in the spectral efficiency (the reader should study and understand this terminology) is not exactly 1 bit/s/Hz on average. In other words, there is a difference between using the average SNR for computing the data rate versus computing the average data rate by averaging the rate obtained across different channel instantiations. The reader should carry out experiments with different values of N_t, observe the variation in the rate obtained, and understand this phenomenon.

We now turn to the underlying mathematics. The beamforming gains (in linear scale) are the eigenvalues of the Wishart matrix. In the MISO and SIMO cases, the Wishart matrix has just one element, which is itself the eigenvalue, i.e., the beamforming gain is

$$\mu = E(\lambda) = \sum_{i=1}^{N} |h_i|^2,$$

where h_i are the elements of the Wishart matrix, and $N = N_t$ or $N = N_r$, as the case may be. Since the expectation of the individual channel gains is unity, i.e.,$\mathbb{E}|h_i|^2 = 1$,

$$\mu := \mathbb{E}(\lambda) = \begin{cases} N_t & \text{for a } 1 \times N_t \text{ MISO system,} \\ N_r & \text{for an } N_r \times 1 \text{ SIMO system.} \end{cases}$$

Since the variance of an exponentially distributed random variable is the square of its mean, and since the $|h_i|$, $1 \leq i \leq N$, are independent,

$$\text{VAR}(\lambda) = \begin{cases} N_t & \text{for a } 1 \times N_t \text{ MISO system,} \\ N_r & \text{for an } N_r \times 1 \text{ SIMO system.} \end{cases}$$

However, the beamforming gains output by NetSim are in dB (log) scale. How does one analytically verify its correctness? The answer lies in Jensen's inequality. Since the log function is concave, Jensen's inequality leads to

$$\mathbb{E}\log_{10}(\lambda) \leq \log_{10}(\mathbb{E}(\lambda)).$$

Here, λ is the eigenvalue of the Wishart matrix and $10\log_{10}\lambda$ is the beamforming gain in the dB scale. Therefore, the beamforming gains (in the dB domain) are bounded as

$$\text{BFGain (dB)} \leq 10\log_{10}(E(\lambda)).$$

But $E(\lambda) = N$, so

$$\text{BFGain (dB)} \leq 10\log_{10} N.$$

In this experiment (and in NetSim), the number of antennas, N, is of the form 2^p, where $p = 0, 1, 2 \ldots$ and therefore, the upper bound on the beam forming gain is

$$\text{BFGain (dB)} \leq 10 \times p\log_{10} 2 \leq 3.01 \times p.$$

4.7 Exercises

1. Quantify the improvement in data rate as a function of N_t (MISO) and N_r (SIMO) in
   a. Low SNR case (SNR $\ll 1$), and
   b. High SNR case (SNR $\gg 1$).
2. (For the Instructor or TA) Assign a set of *personalized* questions that will require each student to run the simulator and generate the results needed to write their reports. For example, different distances between the gNB and UE (which will vary the pathloss), different Tx powers, different ranges for N_t and N_r, etc.

Appendix

We provide the implementation details of this experiment here. These are specific to NetSim v13.3.

Procedure to Import the Workspace for the Experiment

1. Use the following link to download a compressed zip folder that contains the workspace: https://github.com/NetSim-TETCOS/5G_Experiments_v13_2_20/archive/refs/heads/main.zip.
2. Extract the folder.
3. The extracted folder consists of a NetSim workspace file named 5G_advanced_experiments_with_NetSim.netsimexp.
4. Go to the NetSim Home window as shown in Fig. 4.4; go to Your Work and click on Import.
5. In the Import Workspace Window, browse and select downloaded *.netsimexp* file from the extracted directory. Click on the Create a New Workspace option and browse to select a path in your system where you want to set up the workspace folder.
6. Choose a suitable name for the workspace of your choice. Click Import. See Fig. 4.5.
7. The Imported Project workspace will automatically be set as the current workspace.
8. The list of experiments is now loaded onto the selected workspace See Fig. 4.6.

Setting the Network Configurations and Executing the Experiment

To obtain the sample results as given previously, we use the following parameters for network configuration.

1. To set the gNB properties as given in Table 4.1, open the 5G_RAN properties of the gNB-Interface and configure the values as per Table 4.1.

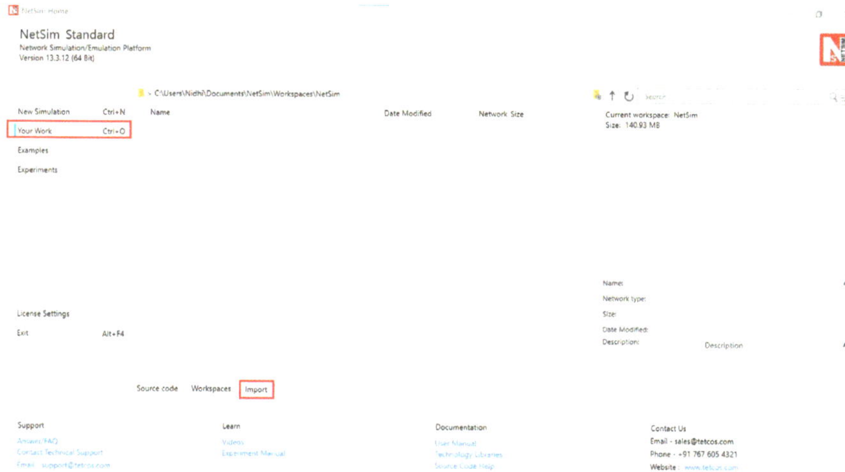

Fig. 4.4 NetSim home page

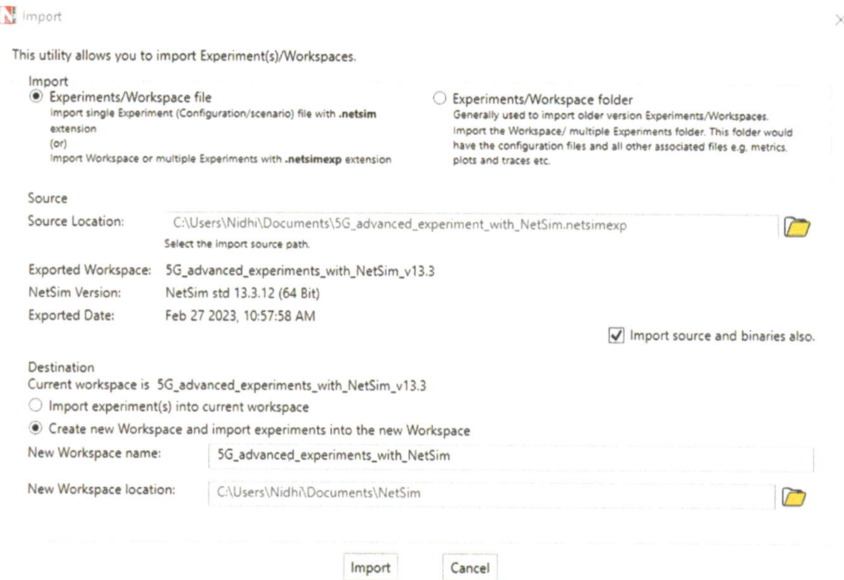

Fig. 4.5 NetSim import workspace window

2. To set the UE properties as given in Table 4.2, open the 5G_RAN properties of the UE-Interface and configure the values as per Table 4.2.
3. The wired link speed is set to 10 Gbps and the Uplink and Downlink BER is set to 0. This eliminates throughput losses in the back-haul links. These are already set in the network configuration.

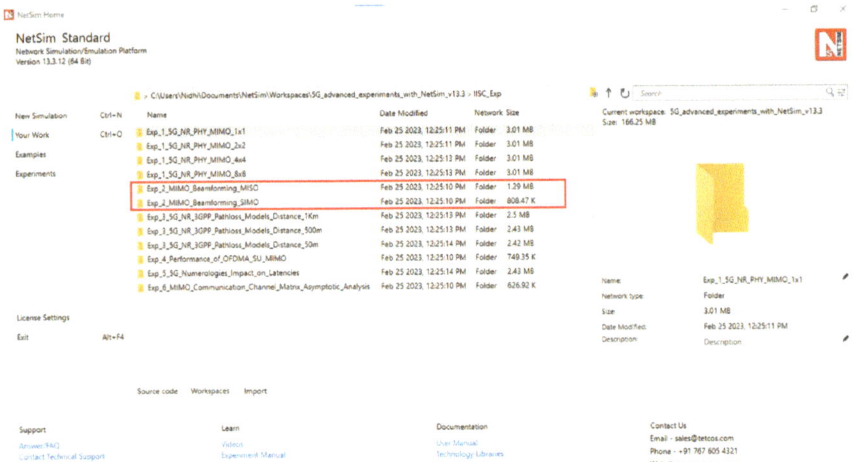

Fig. 4.6 NetSim Your Work Window with the experiment folders inside the workspace

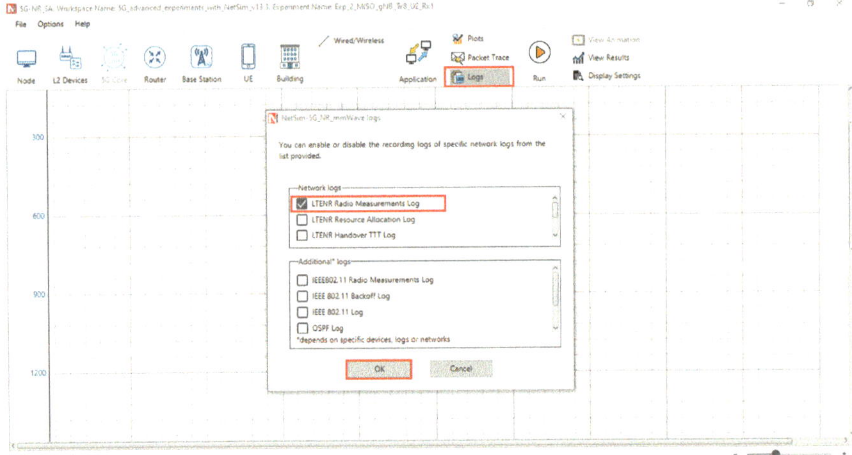

Fig. 4.7 Enabling the LTENR Radio measurement log

4. A downlink CBR application is configured from the wired node to the UE with Transport protocol as UDP, Packet Size of 1460 bytes, and inter-arrival time of 179.69 μs, and the Start Time was set to 1 s.[3]
5. Click on the log icon in the toolbar to enable LTENR Radio measurement, as shown in Fig. 4.7.

[3] The application end time value of 10,000 s is not changed. In NetSim, the application runs for $\min(AppEndTime,\ SimulationTime)$. Since the simulation is run for 10 s, the application runs for only 10 s.

6. For the MISO experiment, vary the Tx Antenna count in the 5G RAN interface of gNB to 1–128.
7. For the SIMO experiment, vary the Rx Antenna count in the 5G RAN interface of UE from 1 to 16.
8. Run the simulation for 10 s, and note down the Application Throughput obtained from the Application Metrics table in the NetSim Results dashboard.
9. Similarly, note down the average Beamforming Gain in dB obtained for the DL application from the log file generated.

Steps to Log the Simulation Results

Steps to calculate the throughput, beamforming gain, SNR, pathloss, and CQI Index:

1. After the simulation, open the NetSim Result dashboard and note down the throughput from the Application Metrics table, as shown in Fig. 4.8.
2. In the results window, expand the Log Files option in the left panel and select LTENR_Radio_Measurements_Log.csv file as shown in Fig. 4.9.
3. This will open the csv file which logs the parameters: beamforming gain, CQI and MCS Indices, pathloss, etc., over time, as shown in Fig. 4.10.
4. Filter the channel to only PDSCH and click on ok since we have considered a DL application from server to UE, as shown in Fig. 4.11.
5. Now select the beamforming gain column and note down the average beamforming gain in dB per layer. See Fig. 4.12.
6. Select the pathloss column and note down the average pathloss value obtained. See Fig. 4.13.
7. In the same way, select the SNR column and note down the average SNR obtained. See Fig. 4.14.

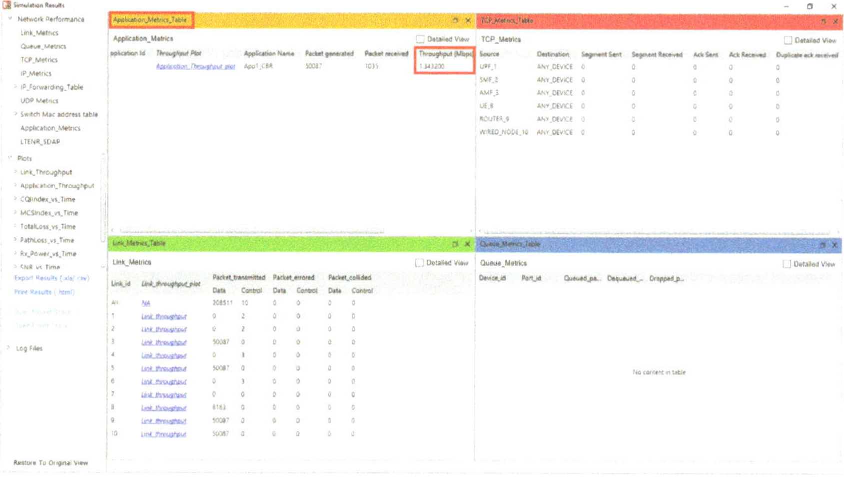

Fig. 4.8 NetSim results window showing application throughput obtained after the simulation

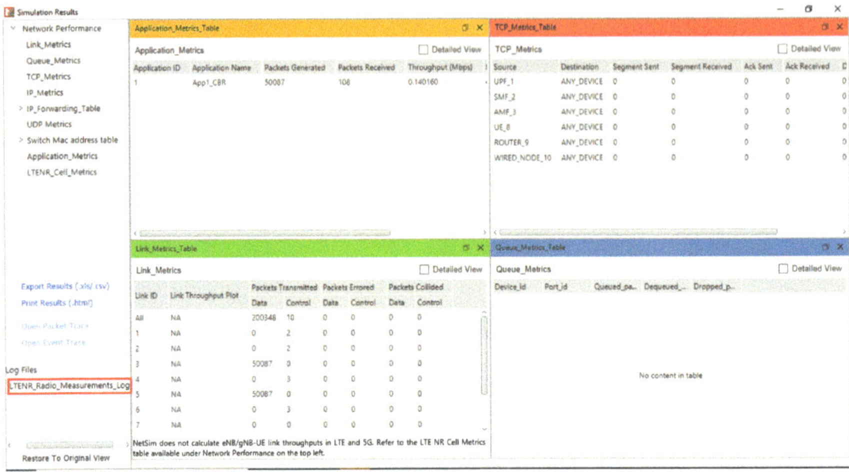

Fig. 4.9 NetSim results window showing access to log file generated

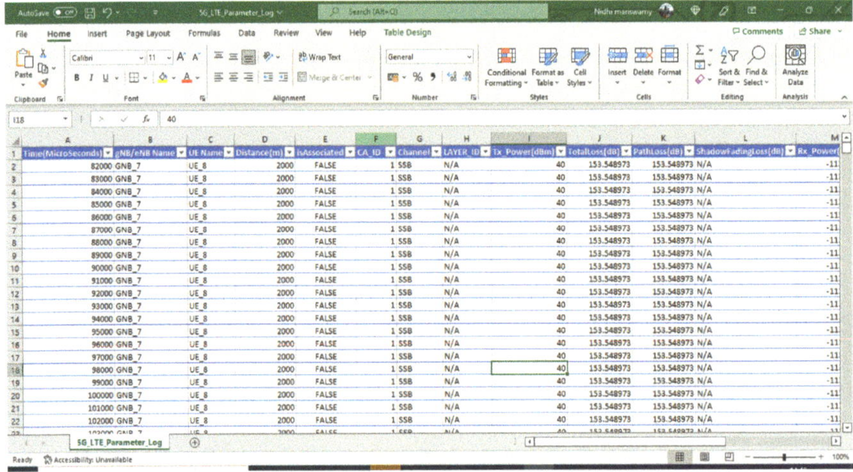

Fig. 4.10 LTENR Radio measurement log file created after simulation

8. Similarly, calculate the average CQI Index and MCS Index. See Figs. 4.15 and 4.16.

Beamforming Gain Plot

1. To generate the beamforming gain plot, open the Radio measurements log file and create a pivot table for this log file by clicking the pivot option at the top of the ribbon under the Insert section, as shown in Fig. 4.17.

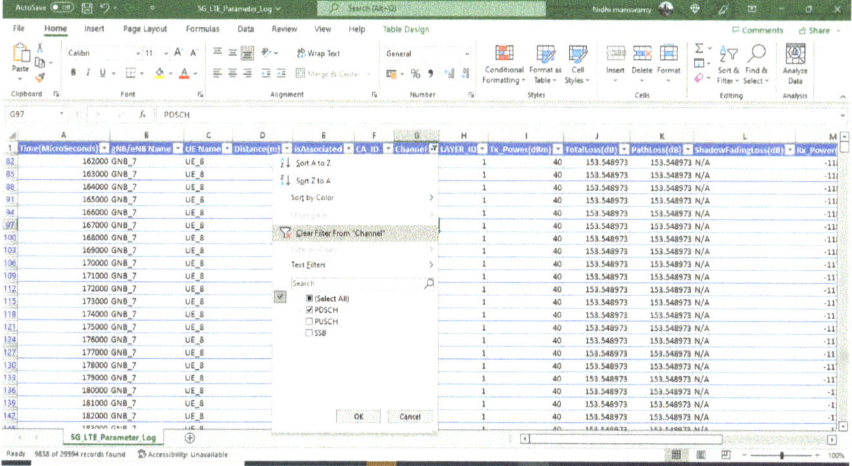

Fig. 4.11 LTENR Radio measurement log file showing the filtering process of the PDSCH/PUSCH column

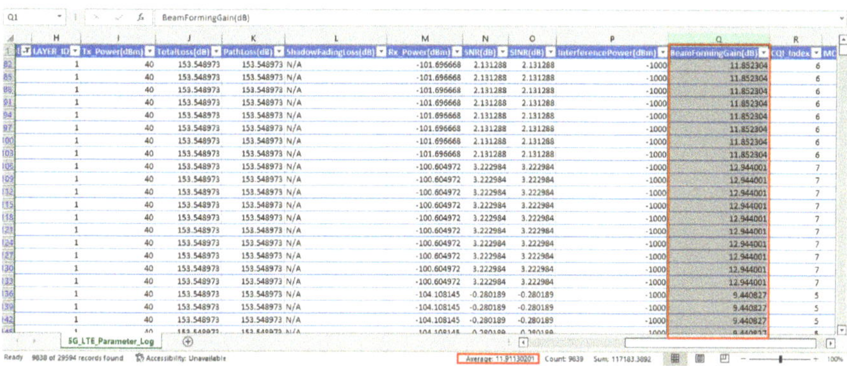

Fig. 4.12 LTENR Radio measurement log file showing average beamforming gain obtained

Fig. 4.13 LTENR Radio measurement log file showing average pathloss obtained

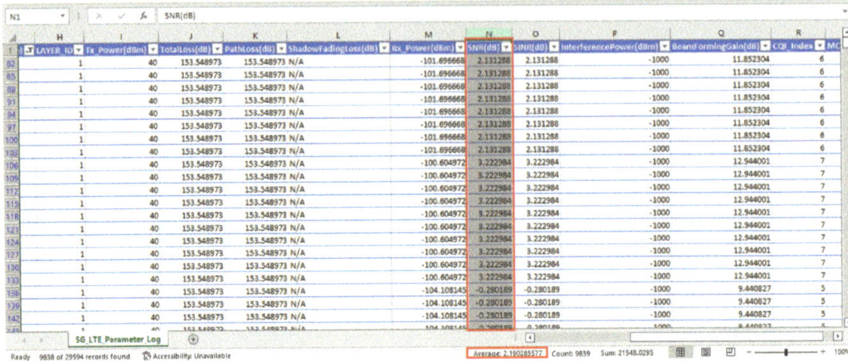

	LAYER_ID	Tx_Power(dBm)	TotalLoss(dB)	PathLoss(dB)	ShadowFadingLoss(dB)	Rx_Power(dBm)	SNR(dB)	SINR(dB)	InterferencePower(dBm)	BeamFormingGain(dB)	CQI_Index
82	1	40	153.548973	153.548973	N/A	-101.696668	2.131288	2.131288	-1000	11.852304	6
85	1	40	153.548973	153.548973	N/A	-101.696668	2.131288	2.131288	-1000	11.852304	6
88	1	40	153.548973	153.548973	N/A	-101.696668	2.131288	2.131288	-1000	11.852304	6
91	1	40	153.548973	153.548973	N/A	-101.696668	2.131288	2.131288	-1000	11.852304	6
94	1	40	153.548973	153.548973	N/A	-101.696668	2.131288	2.131288	-1000	11.852304	6
97	1	40	153.548973	153.548973	N/A	-101.696668	2.131288	2.131288	-1000	11.852304	6
100	1	40	153.548973	153.548973	N/A	-101.696668	2.131288	2.131288	-1000	11.852304	6
103	1	40	153.548973	153.548973	N/A	-101.696668	2.131288	2.131288	-1000	11.852304	6
106	1	40	153.548973	153.548973	N/A	-100.604972	3.222984	3.222984	-1000	12.944001	7
109	1	40	153.548973	153.548973	N/A	-100.604972	3.222984	3.222984	-1000	12.944001	7
112	1	40	153.548973	153.548973	N/A	-100.604972	3.222984	3.222984	-1000	12.944001	7
115	1	40	153.548973	153.548973	N/A	-100.604972	3.222984	3.222984	-1000	12.944001	7
118	1	40	153.548973	153.548973	N/A	-100.604972	3.222984	3.222984	-1000	12.944001	7
121	1	40	153.548973	153.548973	N/A	-100.604972	3.222984	3.222984	-1000	12.944001	7
124	1	40	153.548973	153.548973	N/A	-100.604972	3.222984	3.222984	-1000	12.944001	7
127	1	40	153.548973	153.548973	N/A	-100.604972	3.222984	3.222984	-1000	12.944001	7
130	1	40	153.548973	153.548973	N/A	-100.604972	3.222984	3.222984	-1000	12.944001	7
133	1	40	153.548973	153.548973	N/A	-100.604972	3.222984	3.222984	-1000	12.944001	7
136	1	40	153.548973	153.548973	N/A	-104.108145	-0.280189	-0.280189	-1000	9.440827	5
139	1	40	153.548973	153.548973	N/A	-104.108145	-0.280189	-0.280189	-1000	9.440827	5
142	1	40	153.548973	153.548973	N/A	-104.108145	-0.280189	-0.280189	-1000	9.440827	5

5G_LTE_Parameter_Log — Ready 9838 of 29594 records found — Average: 2.190285577 Count: 9839 Sum: 21548.0295

Fig. 4.14 LTENR Radio measurement log file showing average SNR obtained

	PathLoss(dB)	ShadowFadingLoss(dB)	O2I_Loss(dBm)	Additional_Loss(dB)	Rx_Power(dBm)	SNR(dB)	SINR(dB)	InterferencePower(dBm)	BeamFormingGain(dB)	CQI Index
4	153.548973	N/A	0	0	-118.742437	-14.91448	-14.914481	-1000	-5.193464	0
7	153.548973	N/A	0	0	-118.742437	-14.91448	-14.914481	-1000	-5.193464	0
9	153.548973	N/A	0	0	-118.742437	-14.91448	-14.914481	-1000	-5.193464	0
11	153.548973	N/A	0	0	-118.742437	-14.91448	-14.914481	-1000	-5.193464	0
13	153.548973	N/A	0	0	-118.742437	-14.91448	-14.914481	-1000	-5.193464	0
15	153.548973	N/A	0	0	-118.742437	-14.91448	-14.914481	-1000	-5.193464	0
18	153.548973	N/A	0	0	-118.742437	-14.91448	-14.914481	-1000	-5.193464	0
20	153.548973	N/A	0	0	-118.742437	-14.91448	-14.914481	-1000	-5.193464	0
22	153.548973	N/A	0	0	-118.742437	-14.91448	-14.914481	-1000	-5.193464	0
24	153.548973	N/A	0	0	-118.742437	-14.91448	-14.914481	-1000	-5.193464	0
26	153.548973	N/A	0	0	-117.822231	-13.99428	-13.994275	-1000	-4.273258	0
28	153.548973	N/A	0	0	-117.822231	-13.99428	-13.994275	-1000	-4.273258	0
30	153.548973	N/A	0	0	-117.822231	-13.99428	-13.994275	-1000	-4.273258	0
32	153.548973	N/A	0	0	-117.822231	-13.99428	-13.994275	-1000	-4.273258	0
34	153.548973	N/A	0	0	-117.822231	-13.99428	-13.994275	-1000	-4.273258	0
36	153.548973	N/A	0	0	-117.822231	-13.99428	-13.994275	-1000	-4.273258	0
38	153.548973	N/A	0	0	-117.822231	-13.99428	-13.994275	-1000	-4.273258	0
40	153.548973	N/A	0	0	-117.822231	-13.99428	-13.994275	-1000	-4.273258	0
42	153.548973	N/A	0	0	-117.822231	-13.99428	-13.994275	-1000	-4.273258	0
44	153.548973	N/A	0	0	-117.822231	-13.99428	-13.994275	-1000	-4.273258	0
46	153.548973	N/A	0	0	-116.52915	-12.70119	-12.701194	-1000	-2.980178	0
48	153.548973	N/A	0	0	-116.52915	-12.70119	-12.701194	-1000	-2.980178	0
50	153.548973	N/A	0	0	-116.52915	-12.70119	-12.701194	-1000	-2.980178	0
52	153.548973	N/A	0	0	-116.52915	-12.70119	-12.701194	-1000	-2.980178	0

LTENR_Radio_Measurements_Log — Pivot Table(Custom) — Ready 9840 of 19766 records found — Average: 0.634146341 Count: 9841 Sum: 6240

Fig. 4.15 LTENR Radio measurement log file showing average CQI Index obtained

2. In the pivot table, drop gNB/eNB Name, UE Name, Channel, and Is Associated field under the filters area, drop time (milliseconds) in the Row area, and drop Beamforming gain in the Value area.
3. Set Beamforming gain values to average by clicking the arrow icon present end of the field → value field setting → average, as shown in Fig. 4.18.
4. The final pivot table for average beamforming gain is generated as shown in Fig. 4.19.
5. Copy the values of Row Labels and Average of the Beamforming Gain (dB) header and paste them into another sheet, and generate a plot as shown in Fig. 4.3. This concludes the experiment.

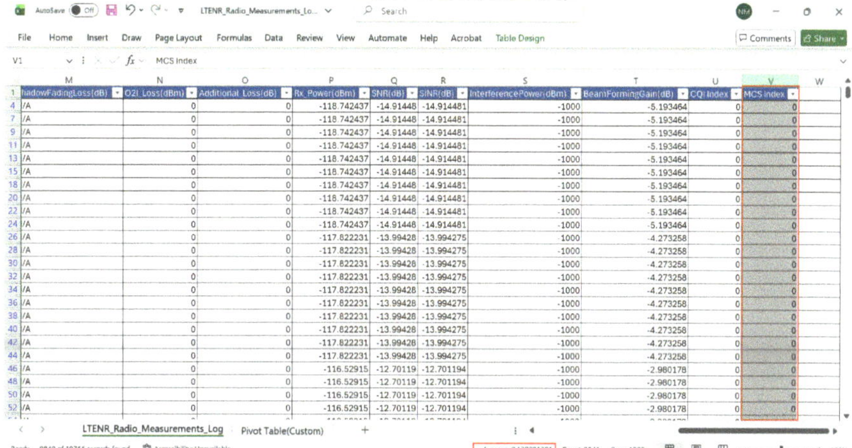

Fig. 4.16 LTENR Radio measurement log file showing average MCS Index obtained

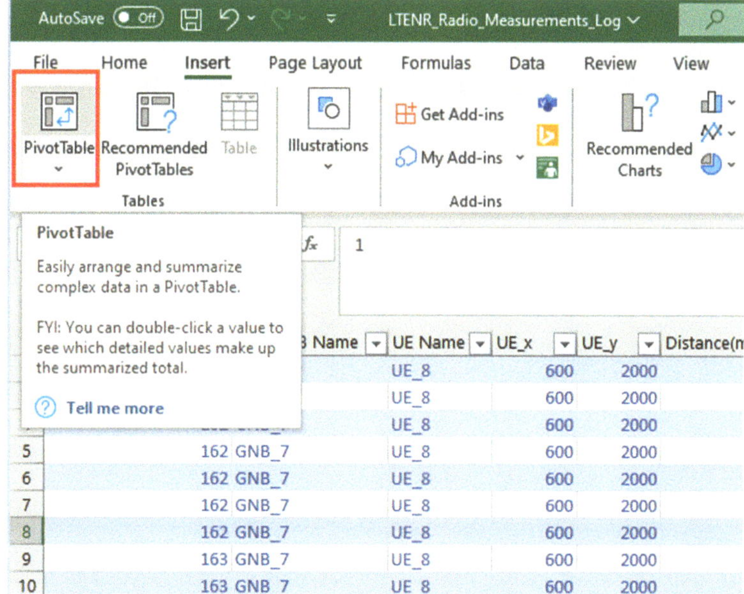

Fig. 4.17 Inserting a pivot table

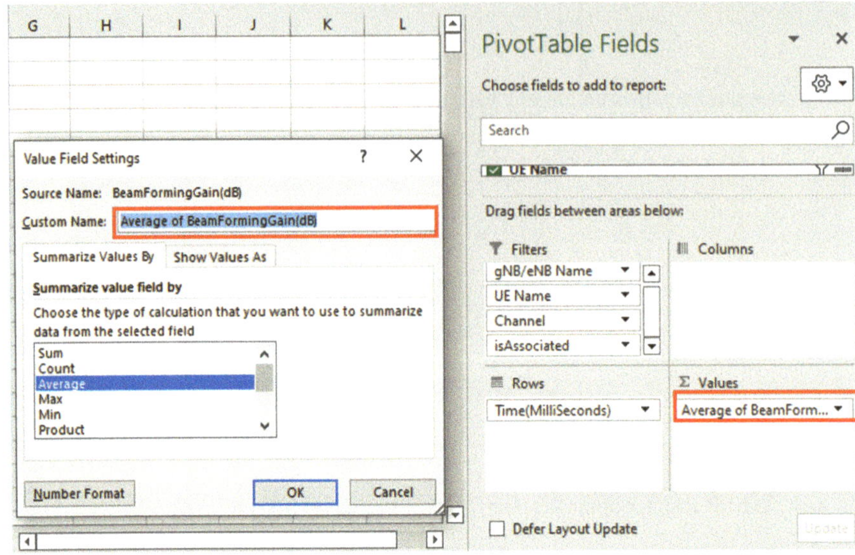

Fig. 4.18 Creating a pivot table

Fig. 4.19 Average beamforming gain for SIMO

A	B
gNB/eNB Name	GNB_7
UE Name	UE_8
Channel	PDSCH
isAssociated	TRUE
Row Labels	Average of BeamFormingGain(dB)
162	-5.193464
163	-5.193464
164	-5.193464
165	-5.193464
166	-5.193464
167	-5.193464
168	-5.193464
169	-5.193464
170	-4.273258
171	-4.273258
172	-4.273258
173	-4.273258
174	-4.273258
175	-4.273258
176	-4.273258
177	-4.273258
178	-4.273258

References

Goldsmith A (2005) Wireless communications. Cambridge University Press

Tse D, Viswanath P (2005) Fundamentals of wireless communication. Cambridge University Press

Chapter 5
Understanding 5G NR Pathloss Models

5.1 Objective

This chapter aims to study the following questions: In the elementary case of 1 gNB communicating with 1 UE over a 5G NR rural network, how does the UE-gNB pathloss vary with the distance between the UE and the gNB and the gNB height? How to choose the optimal height of a gNB and the cell radius while carrying out 5G cell planning in cellular systems? We will demonstrate that the answers to these questions crucially depend on the pathloss of the channel.

5.2 Motivation and Importance

A mobile phone (in the hands of an individual) is the UE; the cell tower is the gNB. Assume the person is in a rural area and in the outdoor scenario. Then, the pathloss would determine the average signal strength displayed on the phone; a higher loss means a lower signal strength. Further, many mobile network operators (e.g., BSNL, Airtel, etc., in India) invest large sums of money in setting up the towers. Then, a natural question is, what should be the tower height[1] so that it gives the users subscribed to the operator the highest signal strength? The answer is not obvious: the higher the height of the gNB, the more likely it is that there exists a line-of-sight path to a given UE, but the signal has to traverse a longer distance, incurring a higher pathloss. The desired cell radius (which depends on the so-called inter-site distance, i.e., the distance between neighboring gNBs) might also play a role here: perhaps a lower height is better for smaller-sized cells, and a greater height is better for large cells. In this experiment, we will understand these trade-offs.

---

[1] The antenna can be placed at different heights on the cell tower. Hence the term "antenna height" would be more appropriate.

L. Yashvanth et al., *Understanding 5G New Radio*, Transactions on Computer Systems and Networks, https://doi.org/10.1007/978-981-92-0112-9_5

Table 5.1 Pathloss equations for rural macro environment for LOS and NLOS states (3GPP 2018)

Scenario	LOS/ NLOS state	Pathloss (dB) (f_c in GHz and d in m)	Shadow fading (σ)	Parameter values and ranges
Rural macro	LOS	$PL_{RMA_{LOS}} = \begin{cases} PL_1, & 10\,\text{m} \le d_{2D} \le d_{BP} \\ PL_2, & d_{BP} \le d_{2D} \le 10\,\text{km} \end{cases}$ $PL_1 = 20\log_{10}(40\pi d_{3D} f_c/3) + \min(0.03h^{1.72}, 10)\log_{10}(d_{3D}) - \min(0.044h^{1.72}, 14.77) + 0.002\log_{10}(h)d_{3D}$ $PL_2 = PL_1(d_{BP}) + 40\log_{10}\left(\frac{d_{3D}}{d_{BP}}\right)$	$\sigma_{SF} =$ 4 $\sigma_{SF} =$ 6	$h_{BS} = 35\,\text{m}$ $h_{UT} = 1.5\,\text{m}$ $W = 20\,\text{m}$ $h = 5\,\text{m}$
	NLOS	$PL_{RMa_{NLOS}} = \max\left(PL_{RMa_{LOS}}, PL_{RMa'_{NLOS}}\right),$ for $10\,\text{m} \le d_{2D} \le 5\,\text{km}$ $PL_{RMa'_{NLOS}} = 161.04 - 7.1 * \log_{10}(W) + 7.5 * \log_{10}(h) - \left(24.37 - 3.7 * \left(\frac{h}{h_{BS}}\right)^2\right) * \log_{10}(h_{BS}) + (43.42 - (3.1 * \log_{10}(h_{BS})) +20 * (\log_{10}(d_{3D}) - 3) * (\log_{10}(f_c)) - (3.2 *(\log_{10}(11.75 * h_{UT}))^2 - 4.97)$	$\sigma_{SF} =$ 8	$5\,\text{m} \le h \le 50\,\text{m}$ $5\,\text{m} \le W \le 50\,\text{m}$ $10\,\text{m} \le h_{BS} \le 150\,\text{m}$ $1\,\text{m} \le h_{UT} \le 10\,\text{m}$

Note

(1) Break point distance[2]: $d_{BP} = 2\pi h_{BS} h_{UT} f_c / c$, where f_c is the centre frequency in Hz, $c = 3.0 \times 10^8$ m/s is the propagation speed of radio waves in free space, and h_{BS} and h_{UT} are the antenna heights at the BS and the UT, respectively. Further h and W denote the average building height and average width of street in the environment, respectively

(2) Further, f_c denotes the centre frequency normalized by 1 GHz, and all distance related values are measured in meters

5.3 The 5G Pathloss Model

The 5G pathloss model is defined in the 3GPP 38.901 standard. For a rural scenario, it is given in Table 5.1.

[2] A question for the reader: Why is it called "break point"? See Molisch (2012) for more details.

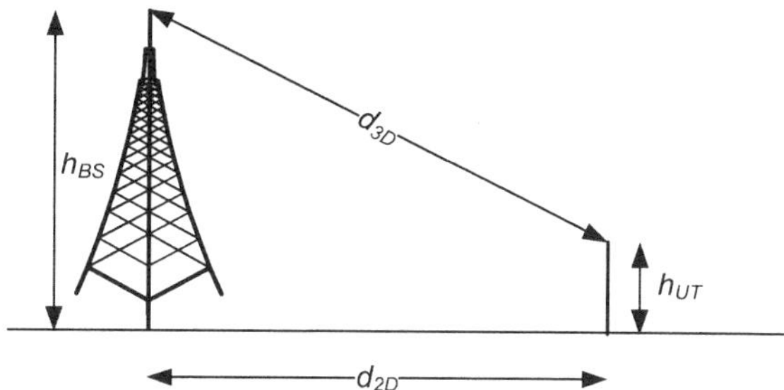

Fig. 5.1 Definition of d_{2D} and d_{3D}

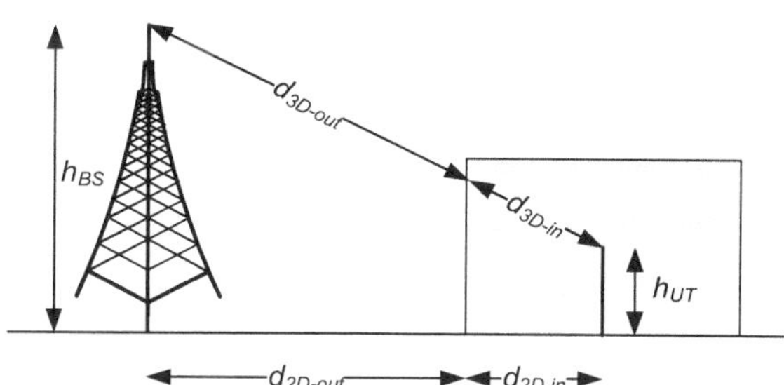

Fig. 5.2 Definition of $d_{\text{2D-out}}$, $d_{\text{2D-in}}$ and $d_{\text{3D-out}}$, $d_{\text{3D-in}}$ for indoor UEs

The distance from the gNB to UE is $d_{\text{3D-out}} + d_{\text{3D-in}} = \sqrt{(d_{\text{2D-out}} + d_{\text{2D-in}})^2 + (h_{\text{BS}} - h_{\text{UT}})^2}$, and the symbols are defined as shown in Figs. 5.1 and 5.2.

Observing the above equation and from Table 5.1, we see that the pathloss is not a simple expression in terms of gNB height. Thus, the other parameters affecting the pathloss are (a) the UE-gNB 2D distance and (b) the UE state: if the UE has a direct path to the gNB without propagating via a scatterer, it is called a line-of-sight (LOS) UE, and if not, it is said to be in the non-LOS (NLOS) state. Consequently, we will investigate a deeper question: how does the UE-gNB pathloss vary for *combinations* of gNB height, UE-gNB 2D distance, and UE states?

5.4 Network Scenario

NetSim UI would display the network topology as shown in Fig. 5.3 when you open the example configuration file.

5.5 Network Configuration

The following settings were configured in the network setup.

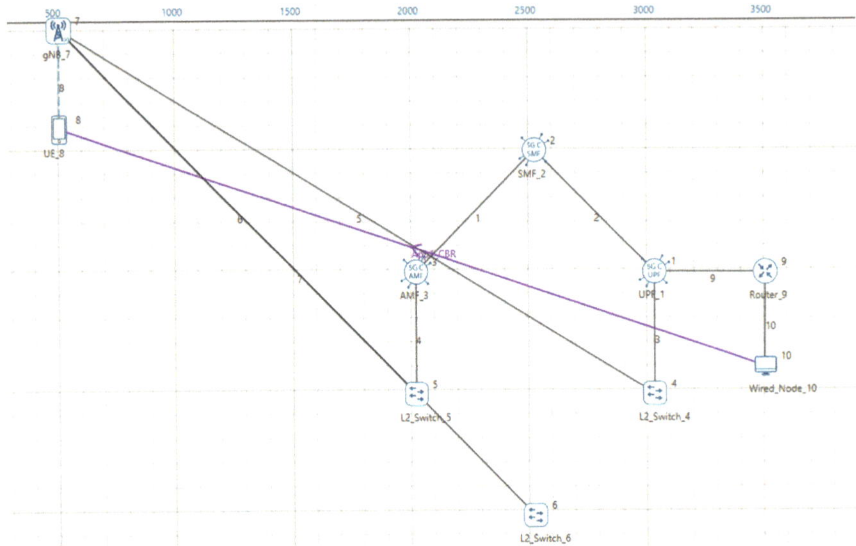

Fig. 5.3 Network topology in this experiment

1. The UE is located at a distance of 50 m from the gNB.
2. The gNB properties are set as per Table 5.2.
3. The UE is configured to have 2 receive and transmit antennas; the other UE properties are similar to the previous experiment.

5.6 Sample Results

We provide the simulation results in Table 5.3, which contain the pathloss values for:

- gNB height varying from 10 to 150 m, in steps of 20 m,
- UE placed at 50, 500, and 1000 m away from gNB, and
- UE states: LOS, NLOS.

5.7 Analytical Verification in Two Cases

In this section, we analytically calculate the pathloss using the 5G pathloss formula from the 3GPP standard (given in Table 5.1) for two cases with common parameters given in Table 5.4, and verify the accuracy of NetSim's output.

Table 5.2 gNB properties

gNB interface 5G RAN	
gNB height (m)	Varied from 10 to 150 m
Tx power (dBm)	40 dBm
Tx antenna count	2
Rx antenna count	2
CA type	Single band
CA configuration	n78
DL: UL ratio	4:1
F_Low (MHz)	3300
F_High (MHz)	3800
Center frequency (MHz)	3550
Numerology	0
Channel bandwidth (MHz)	10
MCS table	QAM64
CQI table	TABLE1
Outdoor scenario	Rural macro
Indoor office type	Mixed office
Pathloss model	3GPPTR38.901-7.4.1
LOS mode	User defined
LOS probability	0 or 1
Shadow fading model	None
Fading and beamforming	No fading
O2I building penetration model	LOW LOSS MODEL

Table 5.3 Pathloss values for various combinations

gNB height (m)	Pathloss (dB)					
	UE 50 m, LOS	UE 50 m, NLOS	UE 500 m, LOS	UE 500 m, NLOS	UE 1 km, LOS	UE 1 km, NLOS
10	77.73	92.39	98.71	132.47	105.57	144.60
30	78.86	84.04	98.72	120.53	105.58	132.21
50	80.58	82.56	98.75	115.30	105.58	126.72
70	82.35	82.72	98.80	111.92	105.60	123.15
90	83.98	83.98	98.86	109.45	105.61	120.51
110	85.44	85.44	98.93	107.52	105.63	118.41
130	86.75	86.75	99.02	105.95	105.66	116.67
150	87.91	87.91	99.11	104.66	105.68	115.20

The gNB heights are shown in column 1. Other columns show the pathloss for different gNB-UE 2D distances (50 m, 500 m, and 1 km) and UE states (LOS/NLOS)

Table 5.4 Various parameters used in the pathloss calculations and their values

Symbol	Description	Value
d_{BP}	Breakpoint Distance	Case dependent
h_{BS}	Height of Base station	10 m
h_{UT}	Height of UE	1.5 m
f_c	Center Frequency in Hz	$3550 * 10^6$ Hz = 3.55 GHz
c	Speed of Light	3×10^8 m/s
W	Street Width	20 m
h	Building Height	5 m

Case 1: gNB Height = 10 m, UE State Is LOS and UE-gNB 2D Distance = 50 m

Breakpoint Distance

- $f_c = \frac{F_{Low}+F_{High}}{2} = \frac{3300+3800}{2} = 3550\,\text{MHz} = 3.55\,\text{GHz}.$
- $d_{Bp} = 2 * \pi * h_{BS} * h_{UT} * (f_c * 1{,}000{,}000{,}000)/c.$
  $= 2 * 3.14 * 10 * 1.5 * \left(\frac{3.55*1{,}000{,}000{,}000}{3*10^8}\right) = 1114.7\,\text{m}.$

Pathloss Calculation

- $d_{2D} = 50\,\text{m},\ d_{3D} = \sqrt{(d_{2D})^2 + (H_{BS} - H_{UT})^2} = \sqrt{(50)^2 + (10 - 1.5)^2} = 50.71\,\text{m}.$

Since $(10 \le d_{2D} \le d_{BP})$,

- $$\text{PL1} = \left(20 * \log 10\left(40 * \text{PI} * \text{distance3D} * f_{c(\text{GHz})}/3\right)\right) + \min((0.03 * \text{pow}(h, 1.72)), 10) * \log 10(\text{distance3D}) \min((0.044 * \text{pow}(h, 1.72)), 14.77) + (0.002 * \log 10(h) * \text{distance3D})$$
  $$= \left(20 * \log_{10}\left(40 * 3.14 * 50.71 * \frac{3.55}{3}\right)\right) + \min((0.03 * \text{pow}(5, 1.72)), 10) * \log_{10}(50.71) - \min((0.044 * \text{pow}(5, 1.72)), 14.77) + \left(0.002 * \log_{10}(5) * 50.71\right) = 77.73\,\text{dB}.$$

Thus, the analytical Pathloss = 77.73 dB, matches the NetSim result.

Case 2: gNB Height = 10 m, UE State Is NLOS and UE-gNB 2D Distance = 50 m

Breakpoint Distance

- $f_c = \frac{F_{Low}+F_{High}}{2} = \frac{3300+3800}{2} = 3550\,\text{MHz} = 3.55\,GHz$
- $d_{Bp} = 2 * \pi * h_{BS} * h_{UT} * (f_c * 1{,}000{,}000{,}000)/c.$
  $= 2 * 3.14 * 10 * 1.5 * \left(\frac{3.55*1{,}000{,}000{,}000}{3*10^8}\right) = 1114.7\,\text{m}.$

Pathloss Calculation

- $d_{2D} = 50\,\text{m}$
- $d_{3D} = \sqrt{(d_{2D})^2 + (H_{BS} - H_{UT})^2} = \sqrt{(50)^2 + (10 - 1.5)^2} = 50.71\,\text{m}.$

  Since $(10 \le d_{2D} \le 5\,\text{km})$, $\text{PL}_{\text{NLOS}} = \max\left(\text{PL}_{\text{LOS}}, \text{PL}'_{\text{NLOS}}\right)$, where

- $$\begin{aligned}\text{PL}'_{\text{NLOS}} &= 161.04 - 7.1 * \log_{10}(W) + 7.5 * \log_{10}(h)\\ &\quad - \left(24.37 - 3.7 * \left(\frac{h}{h_{BS}}\right)^2\right) * \log_{10}(h_{BS})\\ &\quad + \left(43.42 - (3.1 * \log_{10}(h_{BS})\right) * \left(\log_{10}(d_{3D}) - 3\right)\\ &\quad + 20 * \left(\log_{10}(f_c)\right) - \left(3.2 * \left(\log_{10}(11.75 * h_{UT})\right)^2 - 4.97\right)\\ &= 161.04 - \left(7.1 * \log_{10}(20)\right) + 7.5 * \left(\log_{10}(5)\right)\\ &\quad - \left(24.37 - 3.7 * \left(\frac{5}{10}\right)^2\right) * \left(\log_{10}(10)\right)\\ &\quad + \left(43.42 - \left(3.1 * \log_{10}(10)\right)\right.\\ &\quad \left. * \left(\log_{10}(50.71) - 3\right) + 20 * \left(\log_{10}(3.55)\right)\right.\\ &\quad - \left(3.2 * \left(\log_{10}(11.75 * 1.5)\right)^2 - 4.97\right) = 92.39\,\text{dB}\end{aligned}$$

- $$\begin{aligned}\text{PL}_{\text{LOS}} &= \left(20 * \log 10\left(40 * \text{PI} * \text{distance3D} * f_{c(GHz)}/3\right)\right)\\ &\quad + \min((0.03 * \text{pow}(h, 1.72)), 10) * \log 10(\text{distance3D})\\ &\quad - \min((0.044 * \text{pow}(h, 1.72)), 14.77)\\ &\quad + (0.002 * \log 10(h) * \text{distance3D})\\ &= \left(20 * \log_{10}\left(40 * 3.14 * 50.71 * \frac{3.55}{3}\right)\right)\\ &\quad + \min((0.03 * \text{pow}(5, 1.72)), 10) * \log_{10}(50.71)\\ &\quad - \min((0.044 * \text{pow}(5, 1.72)), 14.77)\\ &\quad + \left(0.002 * \log_{10}(5) * 50.71\right) = 77.73\,\text{dB}\end{aligned}$$

$$\text{PL}_{\text{NLOS}} = \max\left(\text{PL}_{\text{LOS}}, \text{PL}'_{\text{NLOS}}\right) = \max(77.73, 92.39)$$

Thus, the analytical pathloss = 92.39 dB, matches the NetSim result.

5.8 Discussion

We now plot the values graphically in Fig. 5.4 and explain the results in these plots from the specifics of pathloss formulas.

- In the LOS plots, the pathloss is flat for different gNB heights when the gNB-UE distance is high, i.e., 500 m and 1 km. When the gNB-UE distance is low, i.e., 50 m, the pathloss increases with gNB height.
- Observe from the LOS pathloss formula that pathloss is proportional to $\log(D_{3D})$. D_{3D} of the 3D distance between the UE and the gNB and is defined as $d_{3D} = \sqrt{(d_{2D})^2 + (h_{BS} - h_{UT})^2}$. It is the hypotenuse of the right triangle, with the base being the gNB-UE 2D distance.
  - Since the length of the hypotenuse is sensitive to the height of the triangle when the gNB-UE distance is small, the pathloss increases with gNB height when the UE is 50 m away, i.e., for smaller distances, the hypotenuse length increases with gNB height.
  - On the other hand, the length of the hypotenuse is almost insensitive to the height of the triangle when the base is much larger than the height. Therefore, when the UE is far, the gNB's height does not have a noticeable impact. Thus, in Fig. 5.4, pathloss is flat when the UE is 500 m and 1 km away.
- We now make the following observations in the NLOS case.

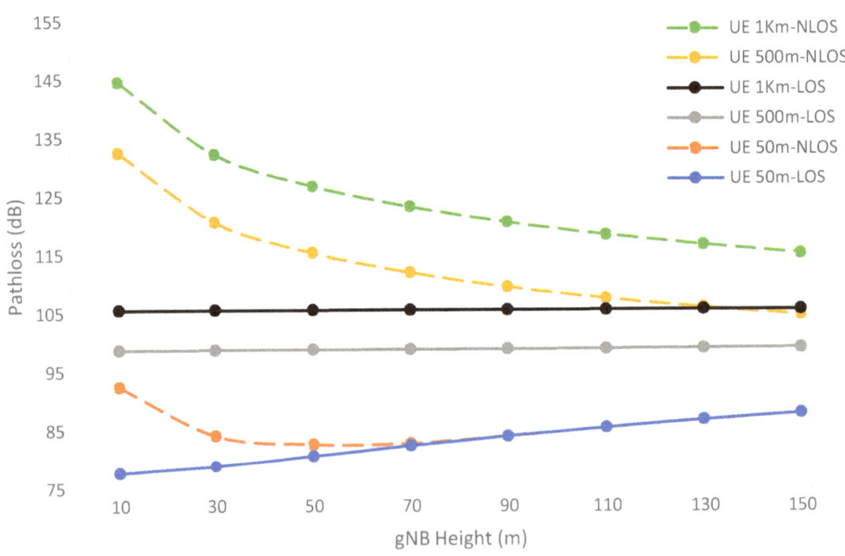

Fig. 5.4 Plot of pathloss versus gNB height for different UE-gNB 2D distances and UE states (LOS, NLOS)

- The NLOS pathloss decreases with gNB height when the gNB-UE distance is high, i.e., 500 m and 1 km. This occurs because, in the NLOS scenario, increasing the gNB height also increases the probability of improved channel strength at the UE.
- Similarly, when the UE is nearer to the gNB, i.e., at 50 m, the NLOS pathloss first decreases and then increases with gNB height. The reason is similar to the above: up to a certain gNB height, the pathloss decreases due to the increased likelihood of witnessing a better (LOS) channel. However, beyond this point, the pathloss increases because of the increased distance between the gNB and UE.
- In fact, these kinds of variations in the pathloss in the NLOS cases also corroborate the pathloss formulas listed in Table 5.1, where we have terms proportional to:
  - $\log(h_{\mathrm{BS}})$, $\log\left((h_{\mathrm{BS}})^2\right)$,
  - $\log(d_{3\mathrm{D}})$, and
  - The reciprocal of $(h_{\mathrm{BS}})^2$.
- Finally, we see that at larger distances, LOS pathloss is almost flat, and the NLOS pathloss decreases as gNB height increases for reasons similar to the ones stated above. Therefore, in the NLOS scenario, an optimal gNB height can be observed. For the example discussed earlier, this optimal height falls within the range of 125–150 m.

5.9 Exercises

1. Make a separate plot with the UE distance on the X-axis and show the behavior for three different values of the gNB height. Recommend the gNB height for different cell radii. Does your recommendation make practical sense?
2. Use MATLAB or Python to plot similar curves from the standard pathloss formulas given in Table 5.1. Compare your results against the NetSim results.
3. (For the Instructor or TA) Generate *personalized* exercises where the student can be asked to
   a. Recommend the gNB height given the cell radius.
   b. Recommend gNB height given the transmit power.
   c. Find the cell radius given the gNB height, transmit power, and noise figure.

An example is as follows:

Suppose, in a rural area, an operator wishes to deploy 5G technology for wireless access. Assume that they can offer cell sizes of radius as sampled from the set {50, 500, 1000} m. Further, they can deploy gNB such that the gNB heights can be as per given in this chapter. They also specify that the minimum average received power for a cell-edge user should at least be − 80 dBm (also known as the *receiver sensitivity*).

Assume that the gNB transmits at 15 dBm power. Further, assume that the only signal propagation loss in the system is the pathloss.

1. Recommend the optimal gNB height so that the minimum target received power is achieved by all users located within a cell of radius 50 m in:

   a. LOS scenario.
   b. NLOS scenario.

   Explain the strategy to arrive at your answer.
2. Recommend the optimal cell radius from the set {50, 500, 1000} so that the minimum target received power is achieved by all users located within a cell having a gNB of height 50 m in

   a. LOS scenario.
   b. NLOS scenario.

   Explain the strategy to arrive at your answer.

(Hint: Understand the implication of the pathloss on the "minimum" received power at a cell-edge UE.)

Appendix

We provide the NetSim implementation details of this experiment in this Appendix. The details are specific to NetSim's v13.3.

Procedure to Import the Workspace for the Experiment

1. Follow the same steps in 1–8 of Appendix in Chap. 4 and get to the stage as shown in Fig. 5.5.

Setting the Network Configurations and Executing the Experiment

1. To set the gNB properties as in Table 5.2, use the 5G RAN interface of the gNB and select the parameters.
2. For UE properties, navigate to the "Physical Layer" properties of the 5G RAN interface of the UE and set Tx Antenna Count = Rx Antenna Count = 2.
3. A downlink CBR application is configured from the Wired Node to UE with a Packet Size of 1460 bytes and inter-arrival time of 1168 μs, and the start time is set to 1 s.
4. Run the simulation for 2 s.
5. Repeat the above for different gNB heights, as desired.
6. Next, for NLOS cases, set the LOS probability to 0 in the 5G RAN properties of the gNB, and run the simulation for various gNB heights.
7. Finally, place the UE at 500, 1000 m away from the gNB and repeat the above procedure.

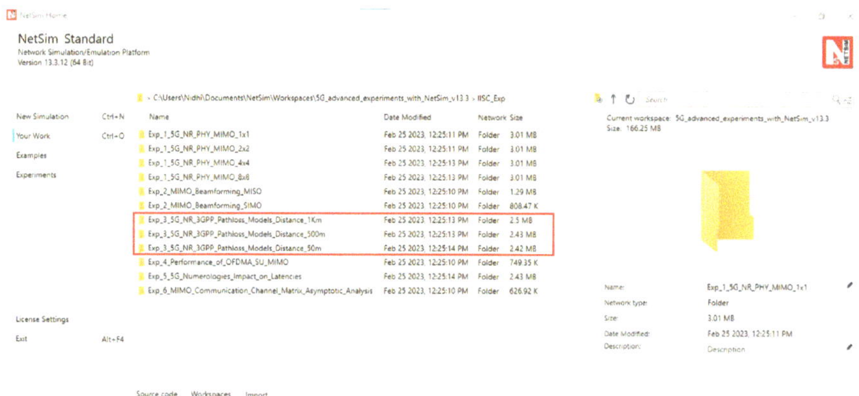

Fig. 5.5 NetSim Your Work Window with the experiment folders inside the workspace

8. Now, for a given simulation run, click on the log icon in the toolbar to enable LTENR Radio measurement, as shown in Fig. 5.6.
9. Then, after the simulation, note down the pathloss from the log file generated for various gNB heights.

Steps to Log the Simulation Results

1. After simulation, open the LTENR Radio measurement Log file from the NetSim result dashboard, as shown in Fig. 5.7.
2. Note down the pathloss value by filtering the channel to PDSCH for each gNB height setting. See Fig. 5.8.

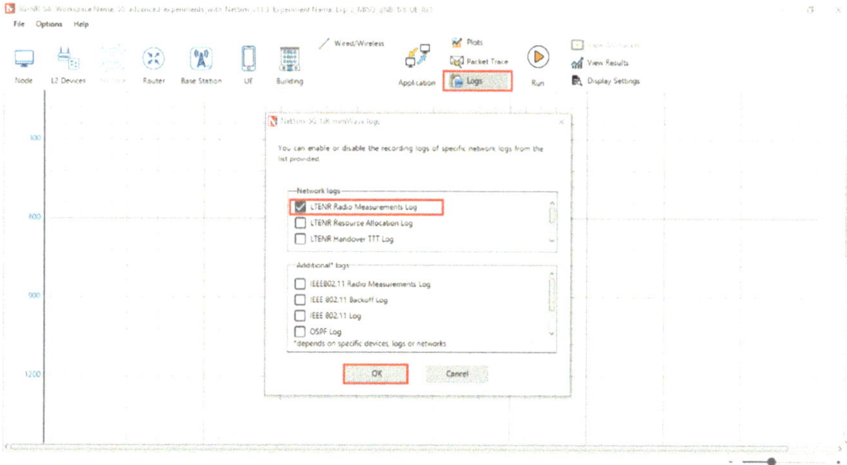

Fig. 5.6 Enabling the LTENR Radio measurement log

Fig. 5.7 NetSim results window

AutoSave Off LTENR_Radio_Measurements_Log Search (Alt+Q) Nidhi manswamy

File Home Insert Page Layout Formulas Data Review View Help Table Design Comments Share

K1 PathLoss(dB)

	Time(MilliSeconds)	gNB/eNB Name	UE Name	Distance(m)	isAssociated	CA_ID	Channel	Layer ID	Tx_Power(dBm)	TotalLoss(dB)	PathLoss(dB)	ShadowFadingLoss(dB)	Rx_Power(dB
4	162	GNB_7	UE_8	50	TRUE	1	PDSCH	1	36.9897	77.734092	77.734092	N/A	-40.
5	162	GNB_7	UE_8	50	TRUE	1	PDSCH	2	36.9897	77.734092	77.734092	N/A	-40.
9	162	GNB_7	UE_8	50	TRUE	1	PDSCH	1	36.9897	77.734092	77.734092	N/A	-40.
10	162	GNB_7	UE_8	50	TRUE	1	PDSCH	2	36.9897	77.734092	77.734092	N/A	-40.
13	163	GNB_7	UE_8	50	TRUE	1	PDSCH	1	36.9897	77.734092	77.734092	N/A	-40.
14	163	GNB_7	UE_8	50	TRUE	1	PDSCH	2	36.9897	77.734092	77.734092	N/A	-40.
17	164	GNB_7	UE_8	50	TRUE	1	PDSCH	1	36.9897	77.734092	77.734092	N/A	-40.
18	164	GNB_7	UE_8	50	TRUE	1	PDSCH	2	36.9897	77.734092	77.734092	N/A	-40.
21	165	GNB_7	UE_8	50	TRUE	1	PDSCH	1	36.9897	77.734092	77.734092	N/A	-40.
22	165	GNB_7	UE_8	50	TRUE	1	PDSCH	2	36.9897	77.734092	77.734092	N/A	-40.
25	166	GNB_7	UE_8	50	TRUE	1	PDSCH	1	36.9897	77.734092	77.734092	N/A	-40.
26	166	GNB_7	UE_8	50	TRUE	1	PDSCH	2	36.9897	77.734092	77.734092	N/A	-40.
30	166	GNB_7	UE_8	50	TRUE	1	PDSCH	1	36.9897	77.734092	77.734092	N/A	-40.
31	166	GNB_7	UE_8	50	TRUE	1	PDSCH	2	36.9897	77.734092	77.734092	N/A	-40.
34	167	GNB_7	UE_8	50	TRUE	1	PDSCH	1	36.9897	77.734092	77.734092	N/A	-40.
35	167	GNB_7	UE_8	50	TRUE	1	PDSCH	2	36.9897	77.734092	77.734092	N/A	-40.
38	168	GNB_7	UE_8	50	TRUE	1	PDSCH	1	36.9897	77.734092	77.734092	N/A	-40.
39	168	GNB_7	UE_8	50	TRUE	1	PDSCH	2	36.9897	77.734092	77.734092	N/A	-40.
42	169	GNB_7	UE_8	50	TRUE	1	PDSCH	1	36.9897	77.734092	77.734092	N/A	-40.
43	169	GNB_7	UE_8	50	TRUE	1	PDSCH	2	36.9897	77.734092	77.734092	N/A	-40.
46	170	GNB_7	UE_8	50	TRUE	1	PDSCH	1	36.9897	77.734092	77.734092	N/A	-40.
47	170	GNB_7	UE_8	50	TRUE	1	PDSCH	2	36.9897	77.734092	77.734092	N/A	-40.
50	171	GNB_7	UE_8	50	TRUE	1	PDSCH	1	36.9897	77.734092	77.734092	N/A	-40.
51	171	GNB_7	UE_8	50	TRUE	1	PDSCH	2	36.9897	77.734092	77.734092	N/A	-40.
54	172	GNB_7	UE_8	50	TRUE	1	PDSCH	1	36.9897	77.734092	77.734092	N/A	-40.
55	172	GNB_7	UE_8	50	TRUE	1	PDSCH	2	36.9897	77.734092	77.734092	N/A	-40.

LTENR_Radio_Measurements_Log Pivot Table(Custom)

Ready 3680 of 7379 records found Accessibility: Unavailable Average: 77.734092 Count: 3681 Sum: 286061.4586 100%

Fig. 5.8 NetSim LTENR Radio measurement log file

References

Molisch AF (2012) Wireless communications, vol 34. Wiley

3GPP (2018) 5G NR: study on channel model for frequencies from 0.5 to 100 GHz. 3GPP TR 38.901 v 15.0.0, Rel 15. ETSI 3rd Generation Partnership Project (3GPP), France

Chapter 6
Understanding 5G NR PHY: Transport Block Processing

6.1 Objective

This experiment has four broad goals:

1. First, to understand the 5G NR physical layer (PHY), i.e., the time–frequency resource grid in the OFDM access scheme of 5G NR.
2. Second, to understand how a packet is transmitted over this OFDM PHY at a transport block level and the assumptions involved in its modeling in NetSim.
3. Third, to analytically compute (as per 3GPP 5G NR standards) the application throughput in a simple use case.
4. Finally, to simulate and analyze the throughput for different PHY parameters (such as the transmit power, MIMO count, and bandwidth) using NetSim v13.3.

6.2 The OFDM Time–Frequency Grid in 5G NR

The 5G NR mandates the use of OFDM as the multiple access scheme for both downlink and uplink transmissions, with variable subcarrier spacing (SCS) to support diverse application scenarios. The smallest physical resource, known as the resource element (RE), comprises one subcarrier and one OFDM symbol.

The time-domain transmission structure comprises frames of duration 10 ms (chosen to facilitate better coexistence with 4G, i.e., LTE). Each frame is composed of 10 subframes of 1 ms each. The 1 ms subframe is then divided into one or more slots in 5G (the definition of a slot in LTE is different; it has exactly two slots in each subframe of duration 1 ms). The slot size depends on the numerology μ (explained in the sequel) and is equal to $\frac{1}{2^{\mu}}$ ms. The number of OFDM symbols per slot is 14 for a configuration using a normal cyclic prefix. When an extended cyclic prefix is used, the number of OFDM symbols per slot is 12. Finally, the data is transmitted over the subcarriers in these OFDM symbols. In Fig. 6.1, we pictorially illustrate the frame

L. Yashvanth et al., *Understanding 5G New Radio*, Transactions on Computer Systems and Networks, https://doi.org/10.1007/978-981-92-0112-9_6

structure when $\mu = 3$, which corresponds to a subcarrier spacing of 120 kHz, and in Fig. 6.2, we illustrate the frame structure for different values of subcarrier spacing.

In the time division duplex (TDD) version, each frame is partitioned into downlink (DL) subframes and uplink (UL) subframes. The downlink subframes are used to send data from the gNB to the UEs, while the uplink subframes are used to send data from the UEs to the gNB. The uplink-downlink ratio is a user-configurable GUI parameter in NetSim. If Internet access is the application of interest, then the

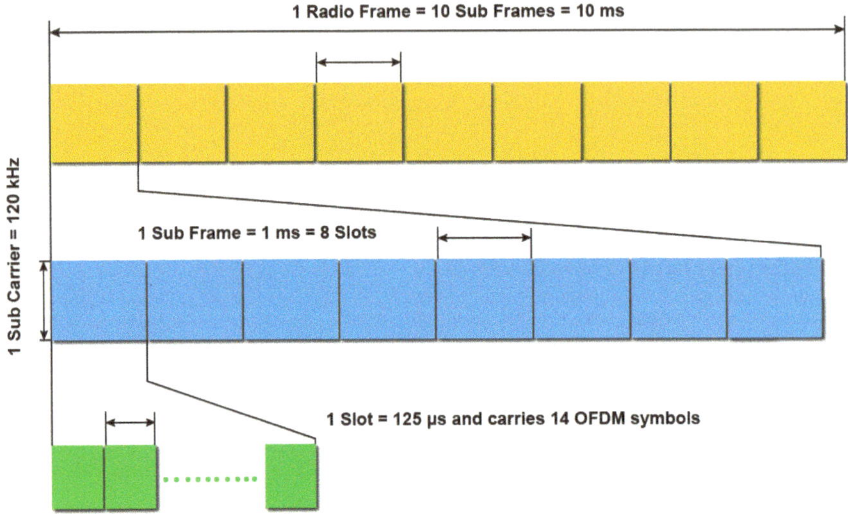

Fig. 6.1 NR frame structure when numerology μ is set to 3

Fig. 6.2 OFDM frame structure. Slot times get shorter as the subcarrier spacing gets larger

downlink part of the frame would usually be substantially larger than the uplink part, due to the asymmetry of Internet access traffic.

In the frequency domain, a group of 12 consecutive subcarriers forms a resource block (RB). 5G NR uses the term numerology to specify the SCS of OFDM; a numerology of μ corresponds to SCS $= 2^{\mu} \times 15$ kHz. The need for the use of variable SCS arises because of the large bandwidths allowed in 5G. In the sub-6 GHz band (FR1), 5G supports a maximum carrier bandwidth of 100 MHz, while in the mmWave band (FR2), the maximum carrier bandwidth is 400 MHz. In order to limit the size of the FFT block used in the OFDM transceiver (which depends on the number of subcarriers) as well as due to somewhat deeper technical reasons related to the delay spread encountered in different frequency bands and the corresponding cyclic prefix duration requirements, 5G uses lower SCS values (e.g., 15, 30, 60 kHz) in FR1 and higher SCS values (e.g., 60, 120, 240 kHz) in FR2.

6.3 Transport Block Processing and Data Transmission in NetSim

- In TDD operation, the UL and DL transmissions are separated in the time domain over different slots/symbols in each frame and occupy the same frequency band. In FDD operation, the UL and DL transmissions happen concurrently but are separated in the frequency domain, with different frequencies used for UL and DL transmissions. In TDD, NetSim does slot-based scheduling. For example, if the DL: UL ratio is 4:1, then the frame is tiled into sets of 5 consecutive slots, and 4 slots are allotted to DL and 1 slot to UL in each tile.
- Higher layer packets arrive at the radio link control (RLC) buffer of each UE for uplink transmission to the gNB it is connected to, and at the RLC buffer of each gNB for downlink transmission to one of the UEs connected to it.
- For downlink transmission, data from the RLC buffers is made available to the MAC scheduler, which then needs to determine which UE to schedule and how much data to transmit to the scheduled UE. To this end, the first step in the MAC scheduler is to determine the transport block size (TBS), which serves as a measure of the amount of data it can send to each UE and therefore helps the scheduler decide which UE to schedule. The TBS is determined based on the channel quality indicator (CQI) reported by the UE using an adaptive modulation and coding (AMC) function. In effect, the TBS ultimately depends on the SNR at the UE, made available to the gNB through the CQI reported by it.
- Now, the SNR observed at the UE is determined by (a) large-scale pathloss and shadowing calculated per the 3GPP's stochastic propagation models (see Chap. 5) and (b) the small-scale fading, which leads to beamforming gains when using MIMO (see Chap. 4). The SNR computations are executed for each associated UE-gNB pair, in DL and UL, at the start of the simulation and again at every

mobility event. In calculating the SNR, the noise power is obtained from its power spectral density: $N_0 = k \times T \times B$. The notation here is explained later.

- In NetSim, the SNR/CQI is not computed using reference signals (as mandated by the standard) but is instead computed on the data channel itself. It is assumed that channel state information (CSI) is instantaneously made available to the transmitter and receiver. This assumption is known as perfect CSIT and CSIR. With perfect CSIT, the transmitter can adapt its transmission rate (MCS) relative to the instantaneous channel state (SNR).
- Thus, based on this SNR, an AMC function is used to compute a wideband CQI value, which indicates the highest modulation and coding scheme (MCS) that the UE can reliably decode if the entire system bandwidth were allocated to it. Further, the MCS is selected from a set of possible values in a standards-specified table. The modulation scheme defines the number of bits that can be carried by a single RE. Modulation schemes supported by 5G include QPSK (2 bits), 16 QAM (4 bits), 64 QAM (6 bits), and 256 QAM (8 bits). The coding scheme determines the code rate used, which is the proportion of bits transmitted that are useful (the remaining are parity bits, used for error correction). It is computed as the ratio of useful bits to total bits that are transmitted. The modulation order Q_m, which denotes the number of bits per RE, and the code rate denoted by R are jointly encoded as the MCS index. These values Q_m and R are then passed to the TBS determination function, as explained in the sequel.
- In a nutshell, the PHY layer in NetSim notifies the MAC about the start of the slot. The MAC sub layer, in turn, seeks a buffer status report from the RLC layer and invokes the MAC scheduler. This then notifies the RLC of the transmission. The RLC then provides the transport block to the MAC, and which is then transferred to the PHY layer. The downlink and uplink data PHY layer channels (referred to as the PDSCH and PUSCH, respectively) receive this transport block as their service data units (SDUs), which are then processed (there are a number of steps in the PHY layer processing, which are not described here) and transmitted over the radio interface.

6.4 Network Scenario

NetSim UI would display the network topology shown in the screenshot in Fig. 6.3 when you open the example configuration file.

6.5 Network Configuration

The following settings are configured in the network setup.

1. The UE is placed 100 m away from the gNB.
2. The gNB is configured with the properties shown in Table 6.1.

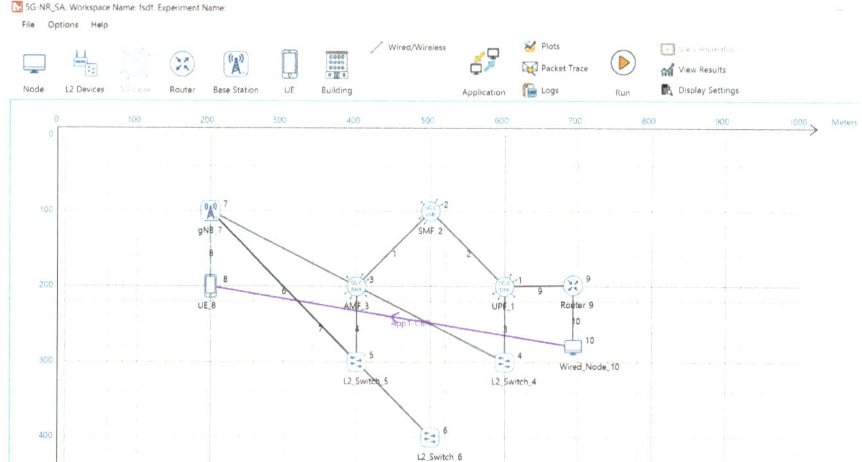

Fig. 6.3 Network topology in this experiment

Table 6.1 gNB properties

gNB interface 5G RAN	
gNB height (m)	10
Tx power (dBm)	31, 34, 37, 40, 43 (varied)
Tx antenna count	1* (varied from 1 to 8)
Rx antenna count	4
CA type	Single band
CA configuration	n261
DL: UL ratio	4:1
Numerology	3
Channel bandwidth (MHz)	50, 100, 200, 400 (varied)
MCS table	QAM256
CQI table	TABLE2
Outdoor scenario	Rural macro
Indoor office type	Mixed office
Pathloss model	3GPPTR38.901-7.4.1
LOS mode	User defined
LOS probability	0
Shadow fading model	None
Fading and beamforming	No fading

3. The number of transmit antennas at the UE is set to 4 (in this example, we consider downlink transmission, so this number does not matter), and the receive antenna counts are varied as 1, 2, 4, and 8 at the UE side through its RAN properties.

6.6 Analytical Estimation of Data Throughput

Steps for Calculating DL Application Throughput

1. The pathloss (dB) is calculated based on the models specified in the 3GPP standards.[1] The large-scale Total Loss is then calculated using the following equation:

$$\text{Large}_{\text{TotalLoss}}\,(\text{dB}) = \text{Pathloss}\,(\text{dB}) + \text{ShadowFading Loss} + \text{O2I Loss}.$$

   Note: O2I Loss in the above refers to the outdoor-to-indoor penetration loss.
2. The received power (per layer) is calculated using the Tx Power (per layer), the Total Loss, and the beamforming gain (per layer) from the following equation:

$$\text{Rx Power}_{\text{Layer}}\,(\text{dBm}) = \text{Tx Power}_{\text{Layer}}\,(\text{dBm}) - \text{Large}_{\text{Total}}\text{Loss}\,(\text{dB}) + \text{BFGain}_{\text{Layer}}\,(\text{dB}).$$

3. Thermal noise power computation:

$$\text{Thermal Noise power} = k \times T \times B,$$

   where k (Boltzmann's constant) $= 1.38 * 10^{-23}$, $T =$ Temperature (in K), $B =$ Bandwidth (in Hz).
4. From the Rx Power and thermal noise power, the SNR is calculated as

$$\text{SNR (Linear)} = \frac{\text{Rx Power (linear)}}{\text{Thermal Noise power (linear)}}.$$

5. From the SNR, the spectral efficiency is calculated as follows (Goldsmith 2005):

$$\text{Spectral Efficiency}_{\text{Layer}} = \log_2(1 + (\text{SNR(Linear)})).$$

6. The CQI index is then looked up from the respective CQI table using the spectral efficiency obtained in the previous step. The CQI table is given below:

   [Format: {*CQI index*, *Modulation scheme*, *code rate* × 1024, *maximum spectral efficiency*}]

---

[1] We do not get into the details of the pathloss computations here. The specifics are explained in Chap. 5.

```
{0,   Modulation_Zero,       0,     0},        //out of range
{1,   Modulation_QPSK,       78,    0.1523},
{2,   Modulation_QPSK,       193,   0.3770},
{3,   Modulation_QPSK,       449,   0.8770},
{4,   Modulation_16_QAM,     378,   1.4766},
{5,   Modulation_16_QAM,     490,   1.9141},
{6,   Modulation_16_QAM,     616,   2.4063},
{7,   Modulation_64_QAM,     466,   2.7305},
{8,   Modulation_64_QAM,     567,   3.3223},
{9,   Modulation_64_QAM,     666,   3.9023},
{10,  Modulation_64_QAM,     772,   4.5234},
{11,  Modulation_64_QAM,     873,   5.1152},
{12,  Modulation_256_QAM,    711,   5.5547},
{13,  Modulation_256_QAM,    797,   6.2266},
{14,  Modulation_256_QAM,    885,   6.9141},
{15,  Modulation_256_QAM,    948,   7.4063},
```

The spectral efficiency in the CQI table corresponding to the chosen CQI index is called the *intermediate spectral efficiency.*

7. The modulation order (Q_m), and its code rate $MCS_{CodeRate}$, together called the MCS index, is extracted from the MCS table, which is given below:

   [Format: {*Index, Modulation order, MCS scheme, MCS-code rate, maximum spectral efficiency*}]

```
{0,   2,  Modulation_QPSK,     120,   0.2344},
{1,   2,  Modulation_QPSK,     193,   0.3770},
{2,   2,  Modulation_QPSK,     308,   0.6016},
{3,   2,  Modulation_QPSK,     449,   0.8770},
{4,   2,  Modulation_QPSK,     602,   1.1758},
{5,   4,  Modulation_16_QAM,   378,   1.4766},
{6,   4,  Modulation_16_QAM,   434,   1.6953},
{7,   4,  Modulation_16_QAM,   490,   1.9141},
{8,   4,  Modulation_16_QAM,   553,   2.1602},
{9,   4,  Modulation_16_QAM,   616,   2.4063},
{10,  4,  Modulation_16_QAM,   658,   2.5703},
{11,  6,  Modulation_64_QAM,   466,   2.7305},
{12,  6,  Modulation_64_QAM,   517,   3.0293},
{13,  6,  Modulation_64_QAM,   567,   3.3223},
{14,  6,  Modulation_64_QAM,   616,   3.6094},
```

```
{15, 6, Modulation_64_QAM, 666, 3.9023},
{16, 6, Modulation_64_QAM, 719, 4.2129},
{17, 6, Modulation_64_QAM, 772, 4.5234},
{18, 6, Modulation_64_QAM, 822, 4.8164},
{19, 6, Modulation_64_QAM, 873, 5.1152},
{20, 8, Modulation_256_QAM, 682.5, 5.3320},
{21, 8, Modulation_256_QAM, 711, 5.5547},
{22, 8, Modulation_256_QAM, 754, 5.8906},
{23, 8, Modulation_256_QAM, 797, 6.2266},
{24, 8, Modulation_256_QAM, 841, 6.5703},
{25, 8, Modulation_256_QAM, 885, 6.9141},
{26, 8, Modulation_256_QAM, 916.5, 7.1602},
{27, 8, Modulation_256_QAM, 948, 7.4063},
```

The modulation order and MCS-code rate are chosen from the MCS table based on the intermediate spectral efficiency obtained above in step 6.

8. The TBS size is then determined using the modulation order and code rate. In 4G LTE, TBS is provided in a table which was a function of the MCS field and the resource block allocation (PRBs). However, 5G NR supports significantly larger bandwidths, together with a wide range of transmission duration, which will result in a large number of tables required. Therefore, 5G NR has opted for a formula-based approach to calculate the TBS. The procedure for TBS determination given in the steps below is as per 3GPP TS 38.214 Section 5.1.3.2 (DL) (3GPP 2018).
9. For calculating the TBS, we must first find the total number of REs available for data transfer (N_{RE})

$$N_{RE} = \min\left(156, N'_{RE}\right) \times n_{PRB},$$

where N'_{RE} = Number of Resource elements per RB available for data transfer, and n_{PRB} = Number of PRBs allotted to the UE. The value N'_{RE} is rounded down to 156 if it is greater than 156, since UE never assumes a resource allocation of more than 156 REs within the bandwidth of a single resource block.

10. The UE determines the no. of REs N'_{RE} that are available for data transfer within the bandwidth of a single resource block using the following formula:

$$N'_{RE} = N_{SC}^{RB} \times N_{Symb}^{Sh} - N_{DMRS}^{PRB} - N_{OH}^{PRB},$$

where
N_{SC}^{RB} = Number of subcarriers in a PRB = 12,
N_{Symb}^{Sh} = Number of Symbols per Slot = 14 (for normal cyclic prefix),

$N_{\text{DMRS}}^{\text{PRB}}$ = Number of REs for demodulation reference signal (DMRS) per PRB in the scheduled duration,

$N_{\text{OH}}^{\text{PRB}}$ = overhead configured by the higher layer parameter: Xoh-PDSCH.

11. N_{RE} is then converted into the number of information bits (N_{info}).

$$N_{\text{info}} = N_{\text{RE}} \times R \times Q_{\text{m}} \quad \text{where} \quad R = \frac{\text{MCS}_{\text{CodeRate}}}{1024}.$$

12. If $N_{\text{info}} \leq 3824$ bits, then TBS is determined based on the 3GPP TS 38.214 Table, and if not, then TBS is determined based on a formula. The detailed steps to determine the TBS are shown in the flowchart given in Fig. 6.4.
13. For $N_{\text{info}} \leq 3824$ bits, N'_{info} is calculated as per the formula given in the flowchart, and then this value is looked up in Fig. 6.5 which is Table 5.1.3.2-1 of 3GPP standard (3GPP 2018). This table comprises 2 columns: index and TBS. The minimum value of the index for which TBS is greater than or equal to the calculated N'_{info} is chosen, and the TBS corresponding to this index is returned as TBS size.

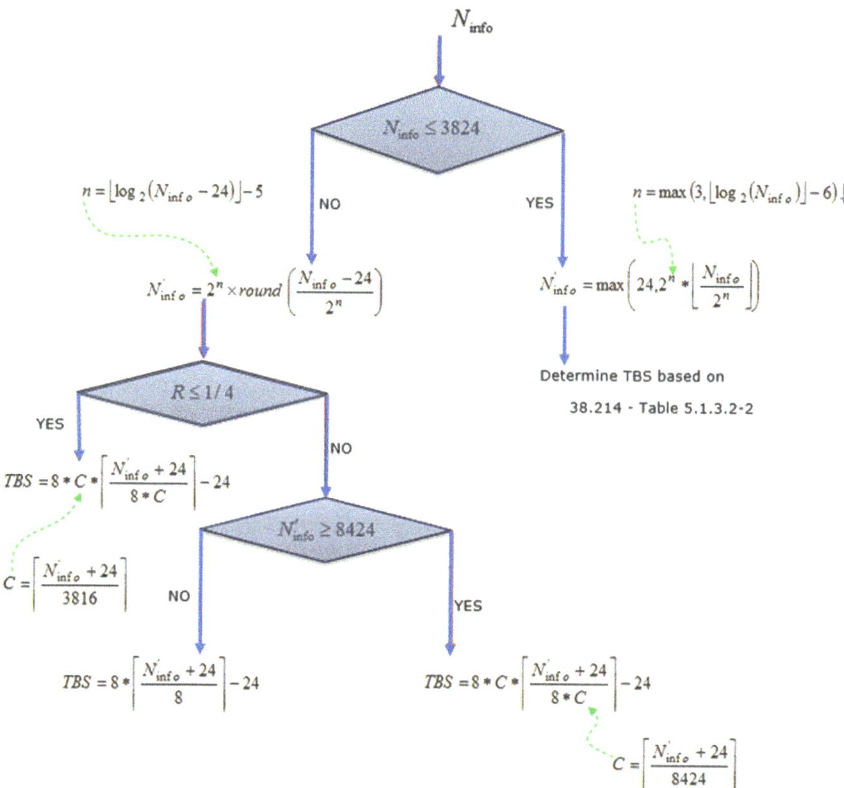

Fig. 6.4 TBS size determination

< 38.214 - Table 5.1.3.2-1: TBS for N_info <= 3824 >

Index	TBS	Index	TBS	Index	TBS	Index	TBS
1	24	31	336	61	1288	91	3624
2	32	32	352	62	1320	92	3752
3	40	33	368	63	1352	93	3824
4	48	34	384	64	1416		
5	56	35	408	65	1480		
6	64	36	432	66	1544		
7	72	37	456	67	1608		
8	80	38	480	68	1672		
9	88	39	504	69	1736		
10	96	40	528	70	1800		
11	104	41	552	71	1864		
12	112	42	576	72	1928		
13	120	43	608	73	2024		
14	128	44	640	74	2088		
15	136	45	672	75	2152		
16	144	46	704	76	2216		
17	152	47	736	77	2280		
18	160	48	768	78	2408		
19	168	49	808	79	2472		
20	176	50	848	80	2536		
21	184	51	888	81	2600		
22	192	52	928	82	2664		
23	208	53	984	83	2728		
24	224	54	1032	84	2792		
25	240	55	1064	85	2856		
26	256	56	1128	86	2976		
27	272	57	1160	87	3104		
28	288	58	1192	88	3240		
29	304	59	1224	89	3368		
30	320	60	1256	90	3496		

Fig. 6.5 TBS calculation from N'_{info}

14. The CRC bits are added prior to channel coding. The threshold of 3824 bits in step (12) is based on the maximum code block size of 3840 bits (3824 bits + 16-bit CRC = 3840 bits), which can be processed by LDPC coding when using 'Base Graph 2'.
15. When $N_{\text{info}} > 3824$, the TBS is found using Fig. 6.4. Further, segmentation may also be required prior to channel coding, and then a 24-bit CRC is added to each segment.
16. If $R \leq \frac{1}{4}$ then LDPC 'Base Graph 2' will be used (maximum code block size of $3816 + 24 = 3840$ bits). If $R > \frac{1}{4}$ and $N'_{\text{info}} > 8424$ bits, then LDPC "Base Graph 1" is used (maximum code block size of $8424 + 24 = 8448$ bits); else, the following equation is used:

$$\text{TBS} = 8 \times \text{ceil}\left(\frac{N'_{\text{info}} + 24}{8}\right) - 24.$$

17. After getting the TBS size, bits per PRB per layer are determined as follows:

    For Layer 1, bits per PRB = bits per PRB (initial) + TBS Size
    For Layer n, bits per PRB_{Ln} = bits per $\text{PRB}_{L(n-1)}$ + TBS Size.

18. Then, the total PRBs available (i.e., the allocated PRB) is

$$\text{Total PRB available} = \text{PRBCount} - \text{ceil}(\text{PRB count} \times \text{OH}_{\text{Downlink}}),$$

    where PRBCount is dependent on bandwidth and μ is shown in the GUI, and we set $\text{OH}_{\text{Downlink}} = 0.18$ as per standard.
19. The slot allocation will then take place, and the bits per slot will be assigned as:

$$\text{bits per Slot} = \text{bits per PRB} \times \text{Total PRB available}.$$

20. Finally, DL Application Throughput = DL MAC Throughput $\times \left(\frac{\text{ApplicationPacketSize}}{\text{MACPacketSize}}\right)$,
    where

$$\text{DL MAC Throughput} = \frac{\text{bits per slot} \times \text{DL fraction}}{\text{slot time}}.$$

Example

We will demonstrate how to derive the throughput for the setting involving 2 × 2 MIMO (2-layers) with 100 MHz Bandwidth and 31 dBm Transmit power. The procedure for TBS determination given in the steps below is as per 3GPP TS 38.214 Section 5.1.3.2 (DL).

1. The pathloss (dB) is calculated based on the pathloss models specified in the 3GPP standards.[2] Pathloss, in this example, turns out to be 122.26 dB.
2. The Total Loss is then calculated using the following equation:

$$\begin{aligned}\text{TotalLoss (dB)} &= \text{Pathloss (dB)} + \text{ShadowFading Loss} \\ &\quad + \text{O2I Loss} + \text{Additional Loss}\end{aligned}$$

    In this example, Shadow Fading Loss = 0, O2I Loss = 0, Additional Loss = 0. Thus, we get,

$$\text{Total Loss} = 122.26 + 0 + 0 + 0 = 122.26\,\text{dB}.$$

---

[2] We do not get into the details of the pathloss computations here. The specifics are explained in Chap. 5.

3. The received power (per layer) is calculated using the Tx Power (per layer), the Total Loss, and the beamforming gain (per layer). Since fading is turned off, the Beamforming (BF) gain per layer is 0 dB.

$$\text{Rx Power}_{\text{Layer}}\ (\text{dBm}) = \text{Tx Power}_{\text{Layer}}\ (\text{dBm}) - \text{Total Loss (dB)} + \text{BFGain}_{\text{Layer}}$$

$$\text{Rx Power}_{\text{Layer 1}} = 27.98 - 122.26 + 0 = -94.28\,\text{dBm},$$

$$\text{Rx Power}_{\text{Layer 2}} = 27.98 - 122.26 + 0 = -94.28\,\text{dBm}.$$

4. Thermal Noise Computation: Thermal Noise = $k \times T \times B$
   k (Boltzmann's constant) $= 1.38 * 10^{-23}$, T (Temperature) $=$ 300 K, $B =$ 100 MHz.
   So, Thermal Noise $= 4.14 \times 10^{-13}\,\text{W} = -93.829\,\text{dBm}$.
5. From the Rx Power and thermal noise, SNR is calculated.
   (a) Rx Power in dBm is converted into mW:

$$\text{Rx Power},\ P = -94.28\,\text{dBm} = 3.73 \times 10^{-10}\,\text{mW}.$$

   (b) Thermal Noise in dBm is converted into mW:

$$\text{Thermal Noise},\ N = -93.82\,\text{dBm} = 4.14 \times 10^{-10}\,\text{mW}.$$

   (c) SNR (Linear) $= \frac{E_b}{N_0} = \frac{\text{Rx Power}}{\text{Thermal Noise}} = \frac{P}{N} = \frac{3.73\times10^{-10}}{4.14\times10^{-10}} = 0.902$.
6. From SNR, the spectral efficiency is calculated as follows:

$$\begin{aligned}\text{Spectral Efficiency}_{\text{Layer}} &= \log_2\left(1 + \left(\frac{E_b}{N_0}\right)\right)\\ &= \log_2(1 + (0.902)) = 0.927.\end{aligned}$$

7. The CQI Index is then looked up from the CQI Table using the spectral efficiency obtained. The table is given below:

```
{0,  Modulation_Zero,     0,   0},        //out of range
{1,  Modulation_QPSK,     78,  0.1523},
{2,  Modulation_QPSK,     193, 0.3770},
{3,  Modulation_QPSK,     449, 0.8770},
{4,  Modulation_16_QAM,   378, 1.4766},
{5,  Modulation_16_QAM,   490, 1.9141},
{6,  Modulation_16_QAM,   616, 2.4063},
{7,  Modulation_64_QAM,   466, 2.7305},
```

```
{8,   Modulation_64_QAM,    567,   3.3223},
{9,   Modulation_64_QAM,    666,   3.9023},
{10,  Modulation_64_QAM,    772,   4.5234},
{11,  Modulation_64_QAM,    873,   5.1152},
{12,  Modulation_256_QAM,   711,   5.5547},
{13,  Modulation_256_QAM,   797,   6.2266},
{14,  Modulation_256_QAM,   885,   6.9141},
{15,  Modulation_256_QAM,   948,   7.4063},
```

Since the Spectral Efficiency is 0.927, from the CQI Table, CQI Index = 3 is chosen.

8. Similarly, the MCS Index is taken from the MCS Table with respect to the spectral efficiency from the CQI Table:

```
{0,   2,  Modulation_QPSK,      120,     0.2344},
{1,   2,  Modulation_QPSK,      193,     0.3770},
{2,   2,  Modulation_QPSK,      308,     0.6016},
{3,   2,  Modulation_QPSK,      449,     0.8770},
{4,   2,  Modulation_QPSK,      602,     1.1758},
{5,   4,  Modulation_16_QAM,    378,     1.4766},
{6,   4,  Modulation_16_QAM,    434,     1.6953},
{7,   4,  Modulation_16_QAM,    490,     1.9141},
{8,   4,  Modulation_16_QAM,    553,     2.1602},
{9,   4,  Modulation_16_QAM,    616,     2.4063},
{10,  4,  Modulation_16_QAM,    658,     2.5703},
{11,  6,  Modulation_64_QAM,    466,     2.7305},
{12,  6,  Modulation_64_QAM,    517,     3.0293},
{13,  6,  Modulation_64_QAM,    567,     3.3223},
{14,  6,  Modulation_64_QAM,    616,     3.6094},
{15,  6,  Modulation_64_QAM,    666,     3.9023},
{16,  6,  Modulation_64_QAM,    719,     4.2129},
{17,  6,  Modulation_64_QAM,    772,     4.5234},
{18,  6,  Modulation_64_QAM,    822,     4.8164},
{19,  6,  Modulation_64_QAM,    873,     5.1152},
{20,  8,  Modulation_256_QAM,   682.5,   5.3320},
{21,  8,  Modulation_256_QAM,   711,     5.5547},
{22,  8,  Modulation_256_QAM,   754,     5.8906},
{23,  8,  Modulation_256_QAM,   797,     6.2266},
{24,  8,  Modulation_256_QAM,   841,     6.5703},
```

{25, 8, Modulation_256_QAM, 885, 6.9141},
{26, 8, Modulation_256_QAM, 916.5, 7.1602},
{27, 8, Modulation_256_QAM, 948, 7.4063},

Since the spectral efficiency is 0.927, MCS Index 3, which corresponds to this spectral efficiency, is chosen, and the Modulation Order = 2.

9. The TBS size is then determined using the modulation order and code rate.
10. Determination of the number of resource elements within the slot

$$N'_{\mathrm{RE}} = N_{\mathrm{SC}}^{\mathrm{RB}} \times N_{\mathrm{Symb}}^{\mathrm{Sh}} - N_{\mathrm{DMRS}}^{\mathrm{PRB}} - N_{\mathrm{OH}}^{\mathrm{PRB}},$$

where $N_{\mathrm{SC}}^{\mathrm{RB}} = 12$ (Number of subcarriers in Physical Resource Block),
$N_{\mathrm{Symb}}^{\mathrm{Sh}} = 14$ (Number of Symbols per Slot),
$N_{\mathrm{DMRS}}^{\mathrm{PRB}} = 0$ (Number of Resource Elements for DMRS per PRB),
$N_{\mathrm{OH}}^{\mathrm{PRB}} = 0$ (PDSCH overhead).
Thus, we have $N'_{\mathrm{RE}} = 12 \times 14 - 0 - 0 = 168$.

11. Total number of resource elements allocated for PDSCH:

$$N_{\mathrm{RE}} = \min\left(156, N'_{\mathrm{RE}}\right) \times n_{\mathrm{PRB}}$$

$n_{\mathrm{PRB}} = 1$ Number of allocated PRBs for the UE,

$$N_{\mathrm{RE}} = \min(156, 168) \times 1 = 156 \times 1 = 156.$$

12. The intermediate number of information bits is calculated as:

$$N_{\mathrm{info}} = N_{\mathrm{RE}} \times R \times Q_{\mathrm{m}}.$$

Now, $R = \frac{\mathrm{MCS}_{\mathrm{CodeRate}}}{1024} = \frac{449}{1024} = 0.438$ and $Q_{\mathrm{m}} = 2$ (Modulation order).
So, $N_{\mathrm{info}} = 156 \times 0.438 \times 2 = 136.65$.

13. Since $N_{\mathrm{info}} \leq 3824$, the TBS size is calculated as

$$N'_{\mathrm{info}} = \max\left(24, 2^n \times \mathrm{floor}\left(\frac{N_{\mathrm{info}}}{2^n}\right)\right),$$

where $n = \max\left(3, \mathrm{floor}\left(\log_2(N_{\mathrm{info}})\right) - 6\right) = \max\left(3, \mathrm{floor}\left(\log_2(136.65)\right) - 6\right) = \max(3, 1) = 3$.
So, $N'_{\mathrm{info}} = \max\left(24, 2^3 \times \mathrm{floor}\left(\frac{136.65}{2^3}\right)\right) = \max(24, 136) = 136$.
Hence, the TBS size will be 136, i.e., index 15 in Fig. 6.5.

14. The bits per PRB per layer are determined based on the TBS size.
15. For Layer 1,

$$\begin{aligned} \mathrm{bitsperPRB} &= \text{bits per PRB (initial)} + \text{TBS Size} \\ &= 0 + 136 = 136. \end{aligned}$$

For Layer 2, $\begin{aligned} \text{bitsperPRB}_{L2} &= \text{bitsperPRB}_{L1} + \text{TBS Size} \\ &= 136 + 136 = 272. \end{aligned}$

16. The total PRBs available is dependent on bandwidth and μ and is shown in the GUI. In this example, PRB available = PRB Count = 66.
17. Then, the total PRB available is calculated as:

$$\text{Total PRB available} = \text{PRBCount} - \text{ceil}(\text{PRB count} \times \text{OH}_{\text{Downlink}}),$$

where PRB Count = 66, $\text{OH}_{\text{Downlink}} = 0.18$ (Per standard).

So, Total PRB available = 66 − ceil(66 × 0.18) = 66 − 11 = 54.

18. Thus, the number of PRBs allocated is 54.
19. The slot allocation will then take place, and the bits per slot will be assigned. Now, bits per Slot = bits per PRB × allocated PRB = 272 × 54 = 14,688 bits = 1836 Bytes, i.e., a slot can transmit a maximum of 1836 Bytes.
20. Throughput estimation: DL UL Ratio = 4 : 1, implies a DL fraction of 0.8. Since $\mu = 3$, the slot time is $\frac{1}{2^3}$ ms. So, DL MAC Throughput $= \frac{1836 \times 8 \times 0.8}{\left(\frac{1}{2^3}\right) \times 10^{-3}} =$ 94 Mbps.

So,

$$\text{DL Application Throughput} = \text{DL MAC Throughput} \times \left(\frac{\text{ApplicationPacketSize}}{\text{MACPacketSize}}\right)$$
$$= 94 \times \left(\frac{1460}{1488}\right) = 92.23\,\text{Mbps}.$$

(This closely matches the NetSim simulation result of 89.81 Mbps, with a slight difference attributed to additional overheads accounted for in the simulations. See Fig. 6.6 entry pertaining to MIMO 2 × 2, Bandwidth 100 MHz, and Tx Power 31 dBm. The entry is marked in boldface green.)

6.7 Sample Results as a Function of System Parameters

In Fig. 6.6, we report the throughputs obtained after running simulations with different antenna counts (MIMO layers), bandwidths, and transmit power values.

In Fig. 6.6, we observe entries marked in:

- *Blue dashed arrows*: When both the bandwidth and the power are doubled, with the MIMO count kept constant, the peak throughput doubles. This is along expected lines: the throughput increases linearly with the bandwidth as long as the SNR is kept constant. When the bandwidth doubles, so does the noise power, so to keep the SNR fixed, the transmit power also needs to double.
- *Red solid arrows*: In the high bandwidth and low power regime, when the bandwidth is doubled with the transmit power and the MIMO count held constant, the peak throughput does not increase but rather decreases. This is because the

MIMO	Total Tx Power[2] (dBm)	Peak Application Throughput (Mbps)			
		Bandwidth (MHz)			
		50	100	200	400
1*1	31	34.98	74.69	89.81	74.69
	34	68.44	100.50	149.38	180.68
	37	96.76	143.25	202.06	298.77
	40	112.18	202.06	287.62	405.23
	43	141.91	234.12	405.23	576.29
2*2	31	71.01	89.81	74.69	84.50
	34	96.76	149.38	180.68	149.38
	37	137.94	202.06	298.77	362.43
	40	194.58	287.62	405.23	597.66
	43	225.42	405.23	576.29	811.52
4*4	31	86.43	74.69	84.50	170.0
	34	143.13	180.68	149.38	170.0
	37	194.58	298.77	362.43	298.77
	40	276.93	405.23	597.66	725.97
	43	390.17	576.29	811.52	1196.44
8*8	31	71.01	84.50	170.0	0
	34	174.26	149.38	170.0	341.05
	37	287.26	362.43	298.77	341.05
	40	390.17	597.66	725.97	597.66
	43	554.91	811.52	1196.44	1452.99

Fig. 6.6 Saturation throughput obtained in the n261 band (gNB-UE distance of 100 m, rural macro pathloss) for various bandwidth-MIMO-TX power combinations. The blue entries show the doubling of throughput when power and BW are both doubled. Red shows examples where throughput decreases with an increase in bandwidth for fixed power and MIMO layers. Green entries are where throughput decreases with an increase in MIMO layers for fixed BW and power

throughput is a concave function of the SNR and is the sum of the throughputs achieved across subcarriers. Doubling the bandwidth while keeping the transmit power fixed halves the SNR per subcarrier, but there are twice as many subcarriers. Due to the concavity of the throughput as a function of the SNR, this leads to a reduction of the overall throughput.

- *Green dotted arrows*: At low power, when the MIMO layers are increased with fixed transmit power and bandwidth, the peak throughput surprisingly decreases.

We explain the above observations in further detail below.

Let us understand the red entries, i.e., throughput of 1×1 MIMO, 31 dBm Tx power for bandwidths of 200 and 400 MHz. We can simplify the PHY rate as equal to $k \times L \times Q \times B \times R$ where k is some constant, L is the number of layers (set to 1 here), Q is the modulation order (2 in this case), R is the code rate, and B is the bandwidth. From Fig. 6.6 (the relevant portion of the figure is reproduced in Table 6.2), we see that when the bandwidth increases, the spectral efficiency decreases because the thermal noise power is proportional to the bandwidths. The received signal power is constant since the transmit power is fixed, thus leading to a reduction in SNR as the bandwidth is increased. Due to the concavity of the relationship between the

Table 6.2 Rx power, noise, SNR, spectral efficiency obtained for different bandwidths at Tx power = 31 dBm

BW (MHz)	Rx power (dB)	Noise (KTB)	SNR	Spectral efficiency (SE)	SE table cut-off	MCS index	MCS code rate	Rate (code rate/ 1024)	Throughput (Mbps)
200	− 91.26	− 90.81	− 0.44	0.927	0.877	3	449	0.438	89.81
400	− 91.26	− 87.80	− 3.45	0.537	0.377	1	193	0.188	74.69

SNR and the throughput, the drop in the MCS[3] due to the reduced spectral efficiency cannot be compensated by the bandwidth increase (i.e., $0.877 \times 200 > 0.377 \times 400$), the net effect is a decline in the throughput (Table 6.2).

Next, we turn to the green entries. We notice that when the MIMO layer count is increased from 4 to 8, the received power (per layer) decreases. This happens because the transmit power is equally divided among all the layers. As SNR reduces, the spectral efficiency per layer decreases. Since the multiplexing gain obtained from multiple MIMO streams does not compensate for the MCS drop (due to lower spectral efficiency): $0.877 \times 4 > 0.377 \times 8$—the consequence is a decrease in throughput. See Table 6.3.

6.8 Exercises

1. Estimate the data throughput analytically for different values of inter gNB-UE distance, transmit power, bandwidth, and MIMO layer count. (Each learner can be given a personalized experiment.)

[3] Refer to Steps 5 through 8 in Sect. 6.6, to understand how spectral efficiency is obtained from the SNR, and how the MCS is set using the spectral efficiency.

Table 6.3 Rx power, noise, SNR, spectral efficiency obtained for each MIMO layer at Tx power = 31 dBm

MIMO layers	BW (MHz)	Rx power (dB)	Noise (KTB)	SNR	Spectral efficiency (SE)	SE table cut off	MCS index	MCS code rate	Rate (code rate/1024)	Throughput (Mbps)
4	50	− 97.28	− 96.83	− 0.44	0.927	0.8770	3	449	0.438	86.43
8	50	− 100.29	− 96.83	− 3.45	0.537	0.3770	1	193	0.188	71.01

Appendix

We provide the implementation details of this experiment here. These are specific to NetSim v13.3.

Procedure to Import the Workspace for the Experiment

1. Follow the same steps in 1–8 of Appendix in Chap. 4 and get to the stage as shown in Fig. 6.7.

Setting the Network Configurations and Executing the Experiment

To obtain the sample results in Fig. 6.6, we use the following parameters for network configuration.

1. To set the gNB properties as given in Table 6.1, open the 5G_RAN properties of the gNB-Interface and configure the values as per Table 6.1.
2. To set the UE properties, open the physical layer properties in the 5G_RAN interface of the UE and configure the values as required.
3. A downlink CBR application was configured from the Wired Node to UE with Packet Size 1460 B and inter-arrival time (varied for each layer count), and the start time was set to 1 s.
4. Run the simulation for 1.2 s.
5. Run the simulation for different MIMO Layers and different Tx powers in gNB and note down the throughput obtained.

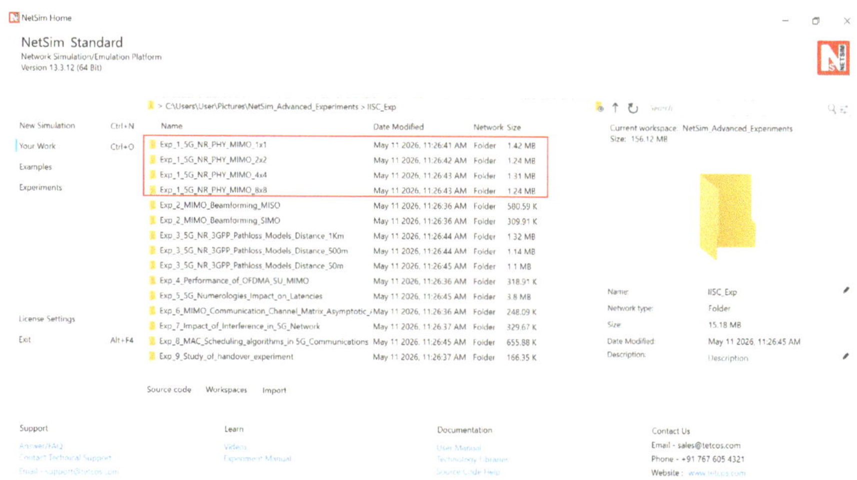

Fig. 6.7 NetSim "Your Work" Window with the experiment folders inside the workspace

Steps to Log the Simulation Results

The results will appear in the Application Metrics table. There is no need for any log files to be opened in this experiment.

Procedure to Record the PRB Count

This is required for computing the analytical throughputs. The first step is to open the NetSim tool and do the following:

1. Open your workspace.
2. Then open the folder containing the configuration files for the first experiment with the MIMO count as per your requirement.
3. Then select the folder with the bandwidth as per your requirement.
4. Now, open any one of the files (it does not matter what the transmit power is!).
   Make sure that you are opening one of the following files, as shown in Fig. 6.8.
5. Then the GUI opens. Right-click on the gNB icon, then click on properties. Select the option "**INTERFACE_4 (5G RAN)**" on the left-hand side. Then scroll down and note down the PRB count. The sample image is shown in Fig. 6.9.

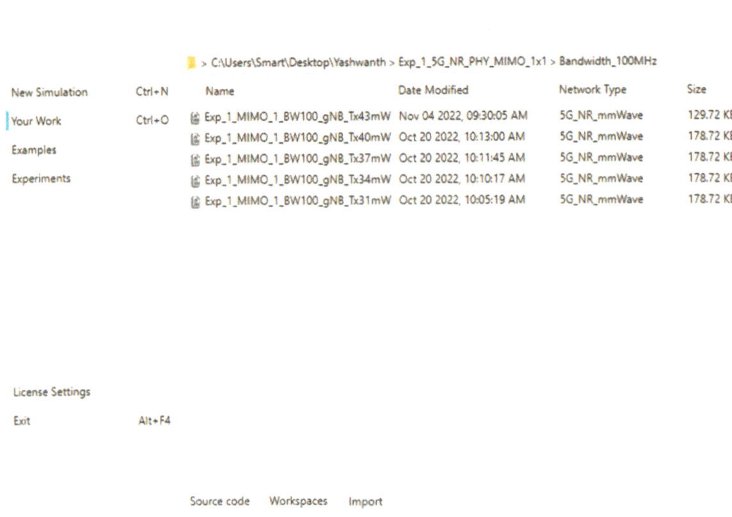

Fig. 6.8 Configuration files of this experiment

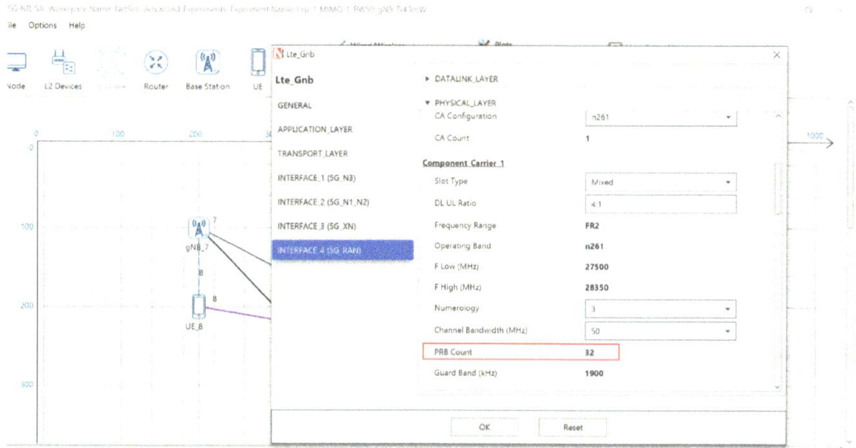

Fig. 6.9 Recording the PRB count

References

3GPP (2018) 5G NR: physical layer procedures for data. 3GPP TS 38.214 v 15.3.0 Rel 15. ETSI 3rd Generation Partnership Project (3GPP), France

Goldsmith A (2005) Wireless communications. Cambridge University Press

Chapter 7
Performance of 5G Single-User MIMO Orthogonal Frequency Division Multiple Access (SU-MIMO-OFDMA)

7.1 Motivation

The 5G cellular system utilizes OFDMA MIMO technology at the physical layer. This technology allows us to exploit the degrees of freedom in frequency, time, and spatial dimensions for multiplexing the data of multiple users. For example, in the downlink direction, i.e., in the data transmission from the gNB to the UEs, the system bandwidth has several OFDM carriers, separated by a specific carrier spacing (for e.g., if 60 kHz subcarrier spacing is used over a bandwidth of 100 MHz, we obtain 132 physical resource blocks (PRBs) with each PRB containing 12 consecutive subcarriers). These PRBs have a time-division slotting (e.g., 0.25 ms), with each slot carrying 14 OFDM symbols, and 18% of these resources are modeled as overheads in NetSim. When such a system is used to transmit data to multiple users on non-overlapping PRBs, it is called OFDMA. In addition, as noted in the previous chapter, MIMO technology exploits spatial multiplexing, which facilitates the transmission of multiple independent streams to different UEs in each OFDM subcarrier. However, if all the spatial streams within an OFDM subcarrier are allotted to only a single user, this technology is called the single-user MIMO or SU-MIMO. In contrast, in Multiuser MIMO (MU-MIMO), different spatial layers within a given subcarrier or PRB can be allotted to different UEs.

In this experiment, we study the performance of OFDMA integrated with SU-MIMO that transmits data to multiple UEs in the downlink direction. In particular, in SU-MIMO, since all the streams on a given PRB are allotted to serve only a single UE, the achievable spatial multiplexing gain at a UE depends on the number of antennas available at the UE, since the base station typically has a larger number of antennas than the UE and can therefore transmit a larger number of layers than what the UE can receive.

L. Yashvanth et al., *Understanding 5G New Radio*, Transactions on Computer Systems and Networks, https://doi.org/10.1007/978-981-92-0112-9_7

7.2 Objective

The goal of this experiment is to simulate and analyze the maximum possible rate of transmission,[1] for the following scenarios using OFDMA configured in a SU-MIMO mode:

- 1 gNB with 8 Tx antennas transmitting to 1 UE with 8 Rx antennas using all the 132 (physical) resource blocks (RBs.)
- 1 gNB with 8 Tx antennas transmitting to 2 UEs, each with 4 Rx antennas and using $\frac{132}{2}$ RBs per UE.
- 1 gNB with 8 Tx antennas transmitting to 4 UEs, each with 2 Rx antennas and using $\frac{132}{4}$ RBs per UE.
- 1 gNB with 8 Tx antennas transmitting to 8 UEs, each with 1 Rx antenna and using $\frac{132}{8}$ RBs per UE.

Subsequently, repeat the simulation for the above scenarios but with the number of UEs now set to 1 for each case, while the antenna counts remain as per the setting of the scenario, and analyze the difference in the transmission rate obtained between a single user and a multi-user-based OFDMA transmission when SU-MIMO is used. Finally, analytically explain the results obtained for different numbers of receive antennas using the eigenvalue properties of the associated Wishart distribution on the channel matrices.

7.3 Introduction

Consider a transmitter with N_t transmit antennas and a receiver with N_r receive antennas. The channel can be represented by the $N_r \times N_t$ matrix $\boldsymbol{H}$ of complex-valued baseband channel gains h_{ij}, representing the gain from the transmit antenna j to the receive antenna i. The $N_r \times 1$ received signal $\boldsymbol{y}$ is given by

$$\boldsymbol{y} = \boldsymbol{Hx} + \boldsymbol{n},$$

where the matrix $\boldsymbol{H}$ is the MIMO channel from the transmitter to the receiver. The MIMO channel can be decomposed into parallel SISO (i.e., non-interfering) streams/channels (Goldsmith 2005). The number of such parallel streams is known as the layer count and is equal to $\min(N_t, N_r)$ under rich scattering conditions. These parallel channels are commonly referred to as the *eigenmodes* of the channel because the singular values of $\boldsymbol{H}$, which are equal to the square root of the eigenvalues of the

[1] We mean the saturation or full buffer case. There is always a packet in the gNB's queue to transmit to every UE; the queue is never empty.

Wishart[2] matrix $W = HH^\dagger$ (for $N_t \geq N_r$), constitute the channel coefficients of the different streams. Specifically, for the layer j, $1 \leq j \leq$ LayerCount, we have

$$y_j = \sqrt{\lambda_j} x_j + w_j,$$

where x_j is the transmitted symbol, λ_j is the corresponding eigenvalue of the Wishart matrix, w_j is the additive white circularly symmetric complex Gaussian noise at the receiver, and y_j is the complex-valued baseband received symbol on the j th eigen mode.

7.4 Network Configuration

1. The properties of gNB and UE are set as per the parameters shown in Tables 7.1 and 7.2, respectively.
2. The wired link properties were configured as shown in Table 7.3.

7.5 Network Scenario

Case 1: 1 gNB—8 Tx antennas, 1 UE—8 Rx antennas

The network scenario for 1 UE with 8 antennas is shown in Fig. 7.1.

Additional Settings

1. The Tx Antenna Count was set to 1, and the Rx Antenna Count was set to 8 in the UE properties.

Case 2: 1 gNB—8 Tx antennas, 2 UEs with 4 Rx antennas each

Network Scenario

The network scenario for 2 UEs, each with 4 antennas, is shown in Fig. 7.2.

Additional Settings

1. The Tx Antenna Count was set to 1, and the Rx Antenna Count was set to 4 in the UE properties of both UEs.

Case 3: 1 gNB 8—Tx antennas, 4 UEs with 2 Rx antennas each

[2] The matrix $\mathbf{W} = \mathbf{HH}^\dagger$ is Wishart distributed when the entries of the channel matrix are complex Gaussian random variables. Further explanation on the Wishart Matrix and it's eigen values is provided in Chap. 4.

Table 7.1 gNB properties

gNB—interface 5G_RAN parameters	
gNB Height	10 m
Tx Power	40 dBm
Duplex mode	TDD
CA Type	SINGLE BAND
CA configuration	n78
DL: UL Ratio	4:1
Numerology	2
Channel bandwidth (MHz)	100
Tx antenna xount	8
Rx antenna count	1
MCS table	QAM256
CQI table	TABLE2
Pathloss model	3GPPTR38.901–7.4.1
Outdoor scenario	Urban Macro
LOS NLOS selection	User defined
LOS probability	1
Shadow fading model	None
Fading and beam forming	RAYLEIGH with EIGEN Beamforming
Coherence time (ms)	10
Additional loss model	None

Table 7.2 UE properties

UE interface 5G RAN	
Tx Power	23 dBm
UE Height	1.5 m
Tx Antenna Count	1
Rx Antenna Count	< varied > as per the experiment needs

Table 7.3 Wired link properties

Wired link parameters	
Wired link speed	10 Gbps
Wired link BER	0
Wired link Propagation Delay	5 μs

Network Scenario

The network scenario for 4 UEs, each with 2 antennas, is shown in Fig. 7.3.

Additional Settings

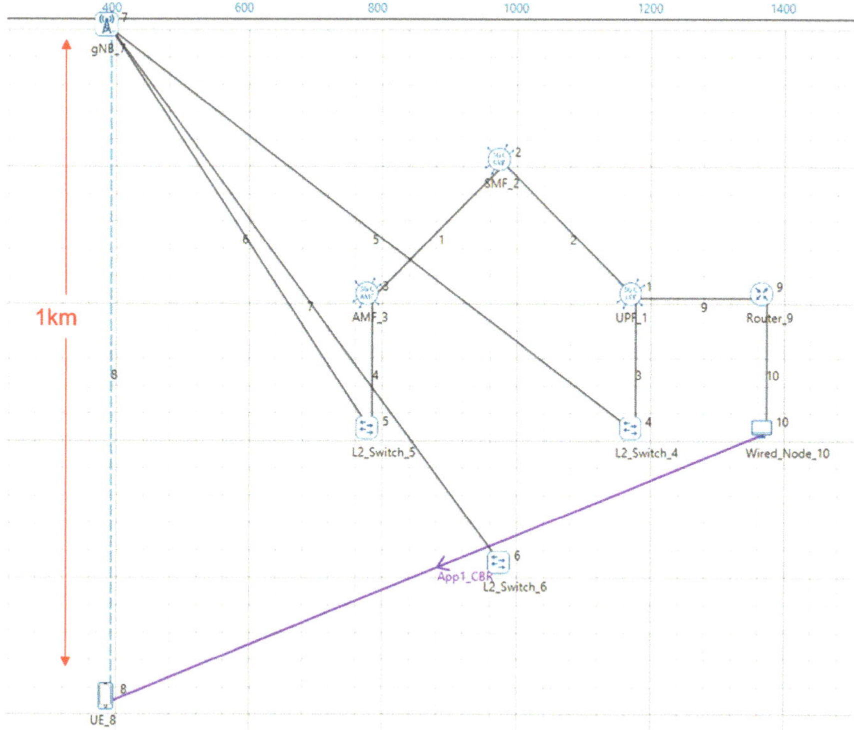

Fig. 7.1 Network topology in this experiment with 1 UE

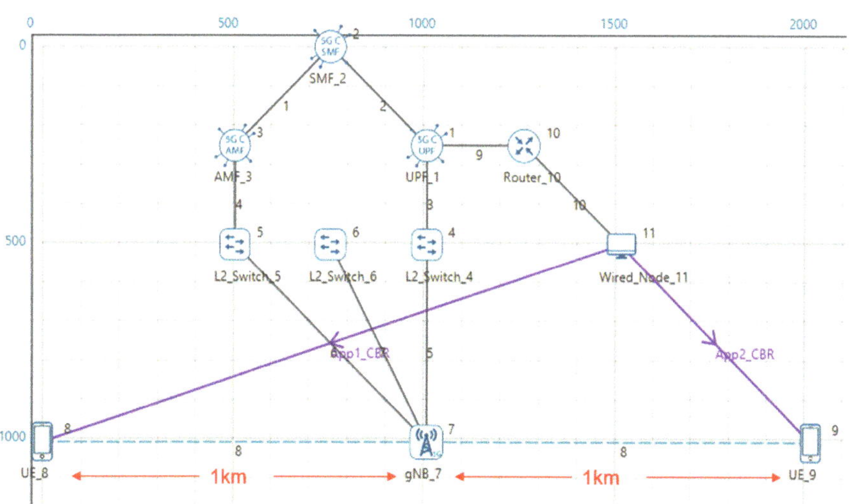

Fig. 7.2 Network topology in this experiment with 2 UEs

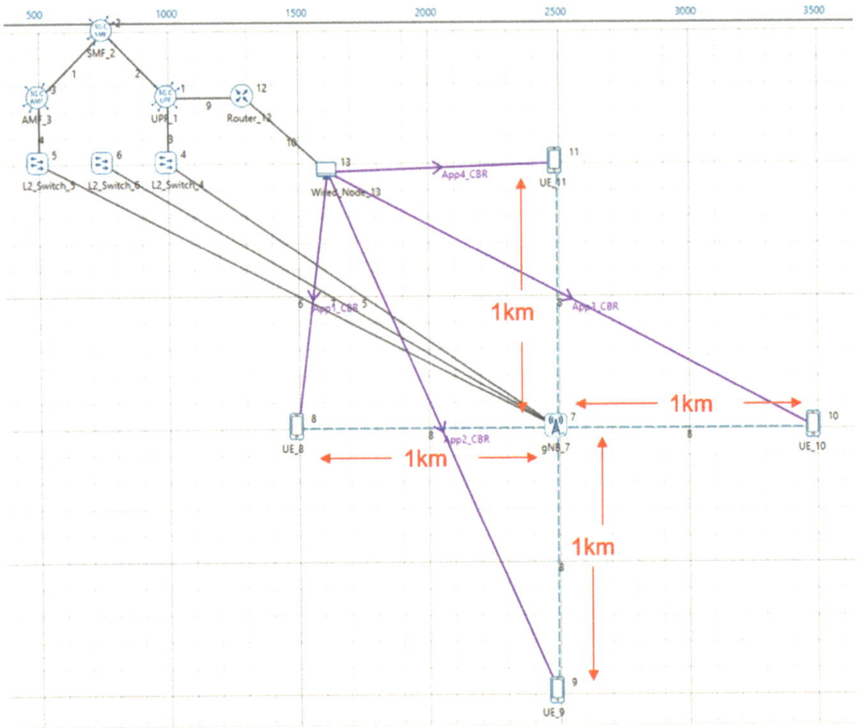

Fig. 7.3 Network topology in this experiment with 4 UEs

1. The Tx Antenna Count was set to 1, and the Rx Antenna Count was set to 2 in the UE properties of all 4 UEs.

Case 4: 1 gNB 8 Tx antennas, 8 UEs with 1 Rx antennas each

Network Scenario

The network scenario for 8 UEs, each with 1 antenna, is shown in Fig. 7.4.

Additional Settings

1. The Tx Antenna Count was set to 1, and Rx Antenna Count was set to 1 in the UE properties of all the 8 UEs.

7.6 Sample Results and Discussions

We combine the results of the four cases and present them in Table 7.4.

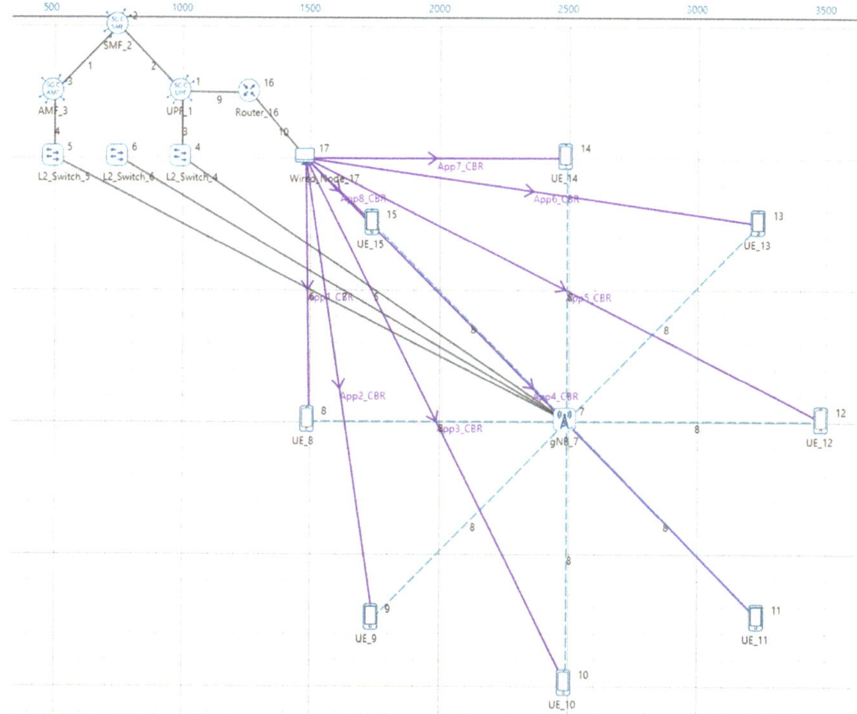

Fig. 7.4 Network topology in this experiment with 8 UEs

Table 7.4 Throughput comparison table for different Rx Antenna Counts

Rx Antenna Count per UE	Throughput (Mbps)	Aggregate Throughput (Mbps)							
	UE_1	UE_2	UE_3	UE_4	UE_5	UE_6	UE_7	UE_8	
8	1909.09	–	–	–	–	–	–	–	1909.09
4	666.92	673.35	–	–	–	–	–	–	1340.28
2	187.81	196.10	190.26	197.27	–	–	–	–	771.464
1	52.20	50.69	52.20	52.09	51.50	50.92	51.39	50.69	411.72

We next compare the aggregate throughputs with single UE peak throughputs, as shown in Table 7.5, which are obtained by allotting the complete bandwidth to this single UE.

From Table 7.5, we observe that even when the bandwidth is shared across the UEs in an OFDMA fashion, the performance is similar to that obtained using a single UE scheduled over the entire bandwidth as long as the number of spatial modes per UE remains fixed (Che and Cheng 2019).

Table 7.5 Throughputs obtained for different Rx Antenna counts in multi and single UE cases

UE Rx antenna count per UE	Aggregate throughput (Mbps)	Single UE peak throughput (Mbps)
8	1909.09 (1 UE 8 Rx Antennas)	1909.09 (1 UE 8 Rx Antennas)
4	1340.28 (2 UEs with 4 Rx Antennas)	1330.46 (1 UE 4 Rx Antennas)
2	771.464 (4 UEs with 2 Rx Antennas)	754.52 (1 UE 2 Rx Antennas)
1	411.72 (8 UEs with 1 Rx Antennas)	414.75 (1 UE 1 Rx Antennas)

7.7 Theoretical Analysis with OFDMA

8 x 8 SU-MIMO

- 2^k carriers in the OFDMA system (we use 2^k for illustrative purposes here; the number of subcarriers can be any multiple of 8 so that it is divisible by the maximum number of users considered in this experiment)
- S : symbol rate per carrier (assuming the same numerology or subcarrier spacing for all carriers)
- L : is the large-scale pathloss, including shadowing (not including the fading). Note that, L is modeled as $c\left(\frac{d}{d_0}\right)^{-\eta}$ where η is the pathloss coefficient.

Then, the expected rate of SU-MIMO data transmission to the single user is given by

$$\mathbb{E}(R_1) = 2^k \times S \times \mathbb{E}\left(\sum_{j=1}^{8} \log\left(1 + \frac{P_j \times L \times \lambda_j}{\sigma^2}\right)\right),$$

where P_j is the power allotted to the jth equivalent (parallel SISO) channel, with $\sum_{j=1}^{8} P_j = P$, the total transmit power. We assume equal power allocation, i.e., $P_j = \frac{P}{8}$, so that

$$\mathbb{E}(R_1) = 2^k \times S \times \mathbb{E}\left(\sum_{j=1}^{8} \log\left(1 + \frac{P \times L \times \lambda_j}{8\sigma^2}\right)\right).$$

8 x 1 SU-MIMO, with 8 such UEs

- 2^{k-3} OFDMA carriers per UE

Then, for the same OFDMA symbol rate, S, and large-scale pathloss, L, the total expected rate for the 8 UEs is given by :

$$\begin{aligned}\mathbb{E}(R_2) &= 2^{k-3} \times S \times \left(\sum_{j=1}^{8} \mathbb{E}\left[\log\left(1 + \frac{P \times L \times \|\mathbf{h}_j\|^2}{\sigma^2} \right) \right] \right) \\ &= 2^k \times S \times \mathbb{E}\left[\log\left(1 + \frac{P \times L}{\sigma^2} \times \|\mathbf{h}_1\|^2 \right) \right].\end{aligned}$$

From matrix theory, we know that the sum of the eigenvalues of a matrix is equal to its trace. Therefore,

$$\sum_{i=1}^{8} \lambda_i = \|h_1\|^2 + \|h_2\|^2 \ldots + \|h_8\|^2.$$

We also know that when the quantities $a, \lambda_1, \lambda_2 > 0$

$$1 + a(\lambda_1 + \lambda_2) \leq (1 + a\lambda_1)(1 + a\lambda_2),$$

and by monotonicity of log function, it follows that

$$\log(1 + a(\lambda_1 + \lambda_2)) \leq \log(1 + a\lambda_1) + \log(1 + a\lambda_2).$$

Extending this inequality inductively, we get

$$\log(1 + a(\lambda_1 + \lambda_2 + \ldots + \lambda_8)) \leq \log(1 + a\lambda_1) + \log(1 + a\lambda_2) + \cdots + \log(1 + a\lambda_8).$$

Using the above expressions, we can compare the expected total downlink throughputs for one 8×8 SU-MIMO UE, with eight 8×1 SU-MIMO UEs, as follows:

$$\begin{aligned}\mathbb{E}[R_1] &= 2^k \times S \times \mathbb{E}\left[\sum_{j=1}^{8} \log\left(1 + \frac{P \times L \times \lambda_j}{8\sigma^2} \right) \right] \\ &\geq 2^k \times S \times \mathbb{E}\left[\log\left(1 + \frac{PL}{8\sigma^2} \times \sum_{j=1}^{8} \lambda_j \right) \right] \\ &= 2^k \times S \times \mathbb{E}\left[\log\left(1 + \frac{PL}{8\sigma^2} \sum_{j=1}^{8} \|\mathbf{h}_j\|^2 \right) \right] \\ &\geq 2^k \times S \times \sum_{j=1}^{8} \frac{1}{8} \mathbb{E}\left[\log\left(1 + \frac{PL}{\sigma^2} \|\mathbf{h}_j\|^2 \right) \right]\end{aligned}$$

$$= 2^k \times S \times \mathbb{E}\left[\log\left(1 + \frac{PL}{\sigma^2}\|\mathbf{h}_1\|^2\right)\right] = \mathbb{E}[R_2]$$

where the first inequality arises from the argument given earlier, and the second equality follows from the equality between the trace of a matrix and the sum of its eigenvalues. The second inequality (third line) follows from the fact that the logarithm function is concave and by an application of Jensen's inequality. We see from Table 7.5 that $\mathbb{E}R_1 = 1909.09$ Mbps, while $\mathbb{E}R_2 = 411.72$ Mbps, showing the large gap between the left- and right-hand sides of the two inequalities in the above argument.

7.8 Exercises

1. Carefully explain your observations. Also, place the UEs at different distances from the gNB and see how the throughput varies. Again, explain your observations.
2. Vary the height of UEs (as this also affects pathloss) and see how the throughput changes. Again, explain your observations.
3. Repeat the above exercises for the asymmetrical location of UEs and infer your observations. In asymmetrical scenarios, the distances between the gNB and each UE are different.

Appendix

We provide the implementation details of this experiment here. These are specific to NetSim v13.3.

Procedure to Import the Workspace for the Experiment

1. Follow the same steps in 1–8 of the Appendix in Chap. 4 and get to the screen as shown in Fig. 7.5.

Performance Evaluation Considerations in NetSim

If fast fading with eigen-beamforming is enabled in NetSim's GUI, then the MIMO link is modeled by parallel SISO channels with the symbol level beamforming gain derived from the eigenvalues[3] of the Wishart matrix.

$$\text{BeamFormingGain (dB)} = 10\log_{10}(\lambda).$$

[3] Note that the eigenvectors are not required as they are only a part of the receive and transmit signal processing; NetSim only needs to work with the equivalent symbol-by-symbol flat fading SISO channels.

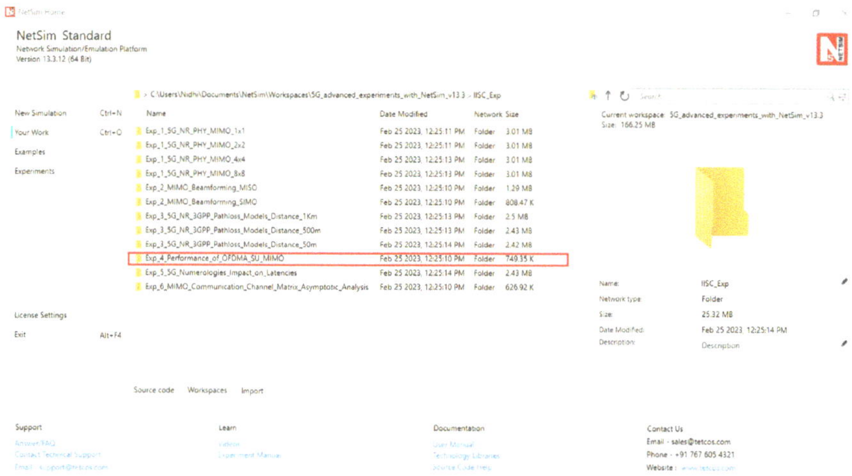

Fig. 7.5 NetSim Your Work Window with the experiment folders inside the workspace

Three assumptions made in NetSim are:

1. Perfect CSIT and CSIR: The channel matrix $\boldsymbol{H}$ is assumed to be known perfectly at the start of each frame, at the transmitter and receiver, respectively. With perfect CSIT the transmitter can adapt its transmission rate (MCS) relative to the instantaneous channel state (SNR).
2. No channel errors: the MCS is chosen based on the channel state, to ensure successful reception of the transmitted data.
3. The transmit power is equally split between all layers transmitted. The justification lies in the fact that at a high SNR, (iterative) water-filling, which is optimal in terms of the sum rate, will lead to nearly equal power allocation across all subcarriers and all layers.

Note that the *LOS probability* parameter in NetSim is solely used to compute the large-scale pathloss per the 3GPP 38.901 standard. This parameter is not used in the channel rank (MIMO layers) computations. The *Fading and Beam Forming* parameter is used to determine (i) the number of MIMO layers and (ii) the gains in each layer, as shown in Table 7.6.

Setting the Network Configurations and Executing the Experiment.

Table 7.6 NetSim modeling of MIMO systems

Parameter drop down option	No. of MIMO layers	Beamforming gain
No fading MIMO unit gain	Min (N_t, N_r)	Unity (0 dB)
No fading MIMO array gain	Min (N_t, N_r)	Max (N_t, N_r)
Rayleigh with eigen beamforming	Min (N_t, N_r)	Eigen values of the wishart matrix

Table 7.7 Application properties with 1 UE as in Case 1

Application parameters: wired Node- UE_8	
Application	CBR
Packet size	1460
Inter packet arrival time (μs)	3.33
Start time	1
Transport protocol	UDP

1. Set the Network configuration as given in Table 7.2, Table 7.3 and Table 7.4 similarly to the procedures explained in the previous chapter.
2. In addition to these properties, we set the following additional "application properties" given in Tables 7.7, 7.8, 7.9, and 7.10 for each of the four cases, respectively.

3. Finally, Run the Simulation for 1.1 s for each of the four scenarios.

Note: While obtaining the single UE peak throughput (2nd column of Table 7.5), just place a single UE in the network and change the receive antenna count as appropriate to the case under study.

Table 7.8 Application properties with 2 UEs as in Case 2

Application parameters		
	Wired Node- UE_8	Wired Node- UE_9
Application	CBR	CBR
Packet size	1460	1460
Inter packet arrival time (μs)	12.97	12.97
Start time	1	1
Transport protocol	UDP	UDP

Table 7.9 Application properties with 4 UEs as in Case 3

Application parameters				
	Wired Node-UE_8	Wired Node-UE_9	Wired Node-UE_10	Wired Node-UE_11
Application	CBR	CBR	CBR	CBR
Packet size	1460	1460	1460	1460
Inter packet arrival time (μs)	53.09	53.09	53.09	53.09
Start time	1	1	1	1
Transport protocol	UDP	UDP	UDP	UDP

Table 7.10 Application properties with 8 UEs as in Case 4

Application parameters								
	Wired Node-UE_8	Wired Node-UE_9	Wired Node-UE_10	Wired Node-UE_11	Wired Node-UE_12	Wired Node-UE_13	Wired Node-UE_14	Wired Node-UE_15
Application	CBR	CBR	CBR	CBR	CBR	CBR	CBR	CBR
Packet size	1460	1460	1460	1460	1460	1460	1460	1460
Inter arrival time (μs)	212.36	212.36	212.36	212.36	212.36	212.36	212.36	212.36
Start time	1	1	1	1	1	1	1	1
Transport protocol	UDP	UDP	UDP	UDP	UDP	UDP	UDP	UDP

Steps to Log the Simulation Results

The results will appear in the Application Metrics table. There is no need for any log files to be opened in this experiment.

References

Che C, Cheng X (2019) Resource allocation for OFDMA systems. Springer International Publishing

Goldsmith A (2005) Wireless communications. Cambridge University Press

Chapter 8
Impact of Numerology on End-to-End Latency and Throughput in 5G NR OFDM Systems

8.1 Objective

5G NR supports flexible numerology with a range of subcarrier spacings (SCSs) by scaling a baseline subcarrier spacing of 15 kHz to support diverse spectrum bands/types and deployment models. The numerology, denoted by μ, can take values from 0 to 4 and specifies the SCS of an OFDM system. In particular, a numerology of μ corresponds to an SCS of $15 \times 2^{\mu}$ kHz and a slot length of $\frac{1}{2^{\mu}}$ ms. Therefore, with μ varying from 0 to 4, SCS varies from 15 to 240 kHz (Zaidi (2016)).

In this experiment, we investigate the impact of numerology on latency and throughput in the following two cases:

- A simple case where one UE is transmitting and receiving the UDP traffic from a server, and
- A complex 5G scenario with Sensors, Cameras, Laptops, and Smartphones having DL and UL, TCP and UDP flows.[1]

8.2 Theory

In NetSim, for data channels, FR1 bands support $\mu = 0, 1, 2$ and FR2 bands support $\mu = 2, 3$. Note that the SCS obtained with $\mu = 0$ corresponds to the SCS used in the LTE (i.e., 4G) system configuration, thereby supporting better coexistence with LTE when LTE and 5G NR are deployed in the same frequency band. In the time domain, the frame length in 5G NR is set to 10 ms, and each frame is composed of 10 subframes of 1 ms each. The 1 ms subframe is then divided into one or more slots in 5G NR, whereas LTE has exactly two slots in each subframe (the definition of a slot is different in LTE and 5G NR, even when the SCS is 15 kHz). The slot

[1] This is adapted from Patriciello et al. (Patriciello et al. (2018)).

L. Yashvanth et al., *Understanding 5G New Radio*, Transactions on Computer Systems and Networks, https://doi.org/10.1007/978-981-92-0112-9_8

size in 5G NR is defined based on μ, and the number of slots is 2^{μ}. The number of OFDM symbols per slot is 14 for a configuration using a normal cyclic prefix. For an extended cyclic prefix configuration, the number of OFDM symbols per slot is 12.

For $\mu = 0$ there is 1 slot per subframe, for $\mu = 1$ there are 2 slots per subframe, for $\mu = 2$ there are 4 slots per subframe, and so on. The number of slots per frame is therefore ten times the number of slots per subframe. Hence for $\mu = 2$, there are 40 slots/frame. We illustrate this in Table 8.1 and pictorially in Fig. 8.1.

Table 8.1 Subcarrier spacing, number of OFDM symbols per slot, slots per frame, and subframe for different numerologies

Numerology	Subcarrier spacing (KHz)	OFDM symbol per slot	Slots per frame	Slots per subframe
0	15	14	10	1
1	30	14	20	2
2	60	14	40	4
3	120	14	80	8
4	240	14	160	16

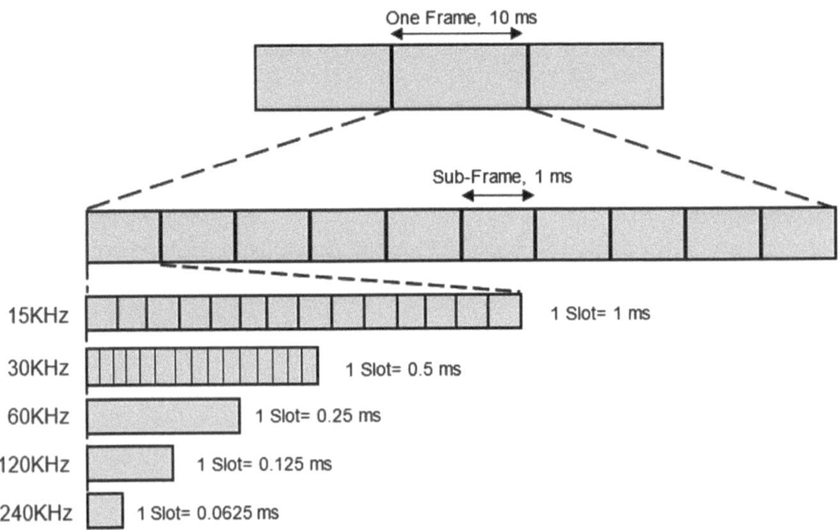

Fig. 8.1 Frame, subframe, and slot structure for different numerologies

8.3 Network Configuration

Case 1: One UE is transmitting and receiving UDP traffic from a server.

This is a simple scenario whereby the UE is transmitting and receiving UDP traffic from a server, as shown in Fig. 8.2.

The UE connects to the gNB, which in turn is connected to the 5G core. The 5G core then connects to the remote server over the cloud (which is represented by the router and WAN links in the Fig. 8.3).

Keeping all other parameters fixed, we vary the numerology μ as 0, 1, and 2 and see its impact on end-to-end latency and application (user) throughput. In terms of application data traffic, the UE has two UDP flows: one for uplink (UL), and another for downlink (DL). These flows are fixed-rate flows. The various simulation parameters related to this experiment are set in NetSim as per the tables given in the Appendix.

Case 2: A complex 5G scenario with Sensors, Cameras, Laptops, and Smartphones having DL and UL, TCP and UDP flows

To model a real-world scenario, we base our simulation on the setup shown in Fig. 8.3. The link between the gNB and the L2_Switches that represent the core network is made with a point-to-point 10 Gb/s link with zero propagation delay. The RAN is implemented using a single gNB, which provides connectivity to multiple UEs over a bandwidth of 100 MHz using a round robin scheduler at the MAC layer. In this case, we have 25 Smartphones, 6 Sensors, and 3 IP Cameras.

In terms of data traffic, the camera (video) and sensor nodes have one UDP flow each that goes in the UL towards a remote node on the Internet. These flows are fixed-rate flows: we have a continuous transmission of 2 Mb/s for the video nodes

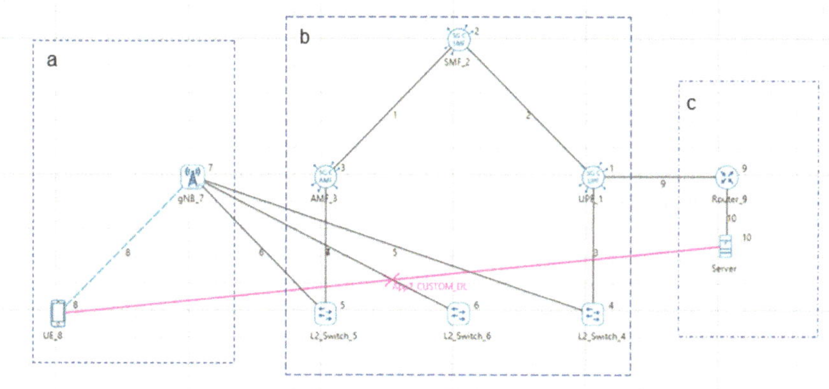

Fig. 8.2 Network scenario. **a** The RAN with UE, **b** 5G Core, and **c** Cloud Server. The device in the RAN has both UL and DL communication with the cloud server

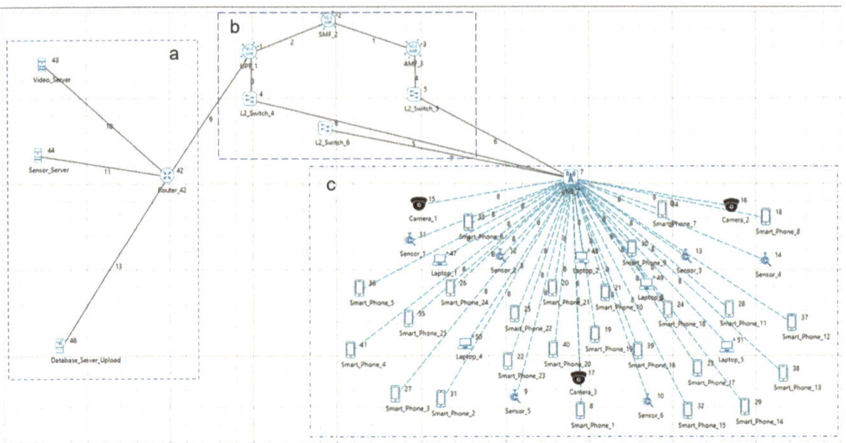

Fig. 8.3 Network scenario. **a** Cloud servers, **b** 5G Core, and **c** The RAN with 25 Smartphones, 6 Sensors, 3 cameras and 6 laptops communicating. The devices in the RAN communicate with respective cloud servers for both Downloads and Uploads

to simulate a 720p24 HD video, and the sensors transmit a payload of 500 bytes every 8 ms, which gives a rate of 0.5 Mb/s. For smartphones, we use TCP as the transmission protocol. These connect to database servers. Each phone downloads a 25MB file and uploads a 1.5MB file. These flows start at different times: the upload starts at a random time between the 4.5th and the 105th simulation seconds, while each download starts at a random time between the 1.5th and the 105th simulation seconds. The laptops download videos at a payload of 1460 bytes every 11.68 ms, which gives a rate of 1 Mb/s. We summarize these settings in Table 8.2.

The numerology μ can take values from 0 to 3 and specifies an SCS of $15 \times 2^{\mu}$ kHz and a slot length of $\frac{1}{2^{\mu}}$ ms. To study the impact of different numerologies and how they affect the end-to-end performance, we measure and analyze the following

Table 8.2 Various parameters of the Traffic flow models for all the devices

	Flows (No of devices)	Traffic rate (Mbps)	Segment/File size (Bytes)	Traffic direction	TCP ACK direction
Camera (UDP)	3	2	500	UL	–
Sensor (UDP)	6	0.5	500	UL	–
Smartphone upload (TCP)	25	–	1,500,000	UL	DL
Smartphone download (TCP)	25	–	25000000	DL	UL
Laptop download (UDP)	5	1	1460	DL	–

Table 8.3 Throughputs obtained for UL and DL UDP flows when numerology is varied from 0 to 2

Application type	Throughput (Mbps)		
	Numerology, $\mu = 0$	Numerology, $\mu = 1$	Numerology, $\mu = 2$
Custom DL (UDP)	5.82	5.82	5.82
Custom UL (UDP)	5.77	5.77	5.77

Table 8.4 Delay obtained for UL and DL UDP applications when numerology is varied from 0 to 2

Application type	Delay (ms)		
	Numerology, $\mu = 0$	Numerology, $\mu = 1$	Numerology, $\mu = 2$
Custom DL (UDP)	2.003	1.139	0.598
Custom UL (UDP)	2.013	1.126	0.598

metrics: a) Throughput of TCP uploads and b) Latency of the UDP uploads and downloads. Different parameters under which this experiment is carried out can be found in Appendix.

8.4 Sample Results

CASE—1

The throughput and delay values are listed in a tabular form in Tables 8.3 and 8.4, respectively, which are also graphically illustrated in Figs. 8.4 and 8.5, respectively.

CASE—2

The throughput and delay values under this case are listed in a tabular form in Tables 8.5 and 8.6, respectively. Further, for better illustration, we provide a graphical representation of the values for throughput and delay in Camera video and Sensor UL applications in Figs. 8.6 and 8.7, respectively, that of the Laptop video DL application in Fig. 8.8, and that of the Smartphone UL application in Fig. 8.9.

8.5 Discussions

For UDP applications, the numerology μ does not impact the throughput. This can be explained as follows:

In UDP, the gNB sends data at a fixed rate, i.e., with a fixed number of packets, each of a fixed size, per second. If this fixed rate can be accommodated within the

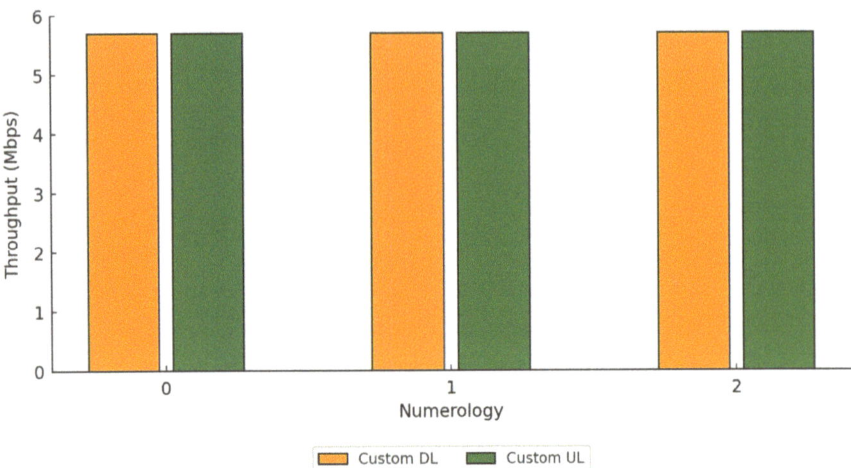

Fig. 8.4 Custom DL and UL throughput versus numerology. Numerology has no impact on throughput

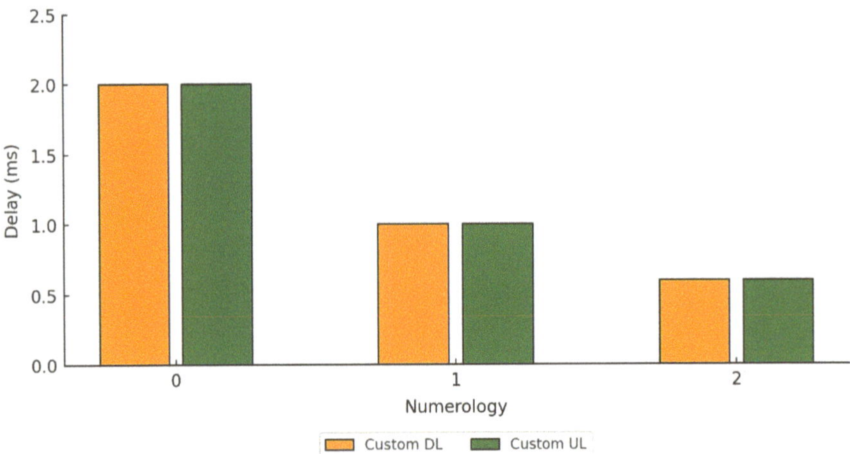

Fig. 8.5 Custom DL and UL delay versus numerology. The delay for both DL and UL decreases as the numerology is increased

available time–frequency resources of an OFDM frame in (say) $\mu = 0$, then it can be accommodated within one frame at all MCSs since the total resources available in the OFDM resource grid are the same at all SCSs. This is because, as μ increases, although the slot duration reduces, resulting in double the number of slots within each frame, the SCS also doubles, which halves the number of REs within the bandwidth. Thus, the total number of time–frequency resources available in a frame is fixed for all values of μ. As a consequence, the data sent over UDP continues to be received

Table 8.5 Average and aggregate throughputs obtained for camera, sensors, smartphones, and laptops, when numerology is varied from 0 to 2

Application type	Average throughput (Mbps)		
	Numerology, $\mu = 0$	Numerology, $\mu = 1$	Numerology, $\mu = 2$
Smartphone DL (TCP)	80.44	151.82	214.34
Smartphone UL (TCP)	73.60	147.10	290.84
Sensor UL (UDP)	0.49	0.49	0.49
Camera video UL (UDP)	1.99	1.99	1.99
Laptop video DL (UDP)	0.99	0.99	0.99

Table 8.6 Average delay obtained for cameras, sensors, smartphones, and laptops when numerology is varied from 0 to 2

Application type	Average delay (ms)		
	Numerology, $\mu = 0$	Numerology, $\mu = 1$	Numerology, $\mu = 2$
Smartphone DL (TCP)	1238.84	658.00	433.60
Smartphone UL (TCP)	91.93	46.18	23.45
Sensor UL (UDP)	2.34	1.17	0.58
Camera video UL (UDP)	2.35	1.17	0.59
Laptop video DL (UDP)	2.52	1.26	0.76

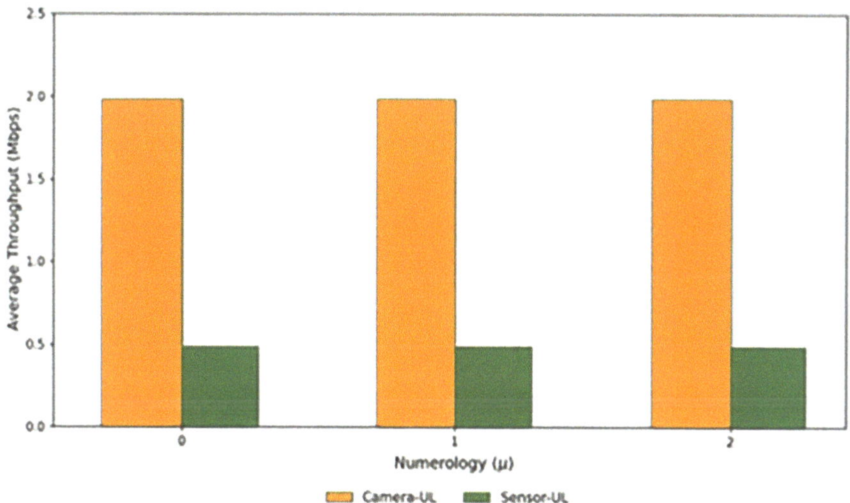

Fig. 8.6 Average uplink throughputs for Cameras and Sensors remain the same as the numerology is increased. This is because the flow is UDP

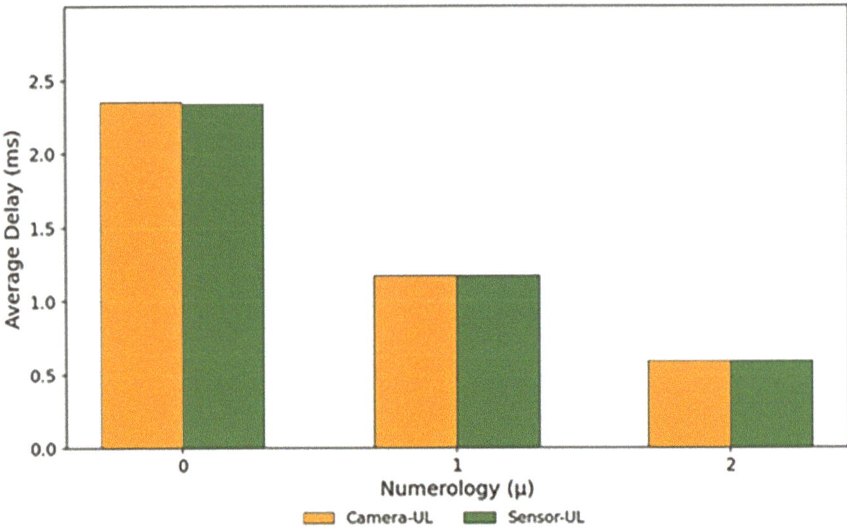

Fig. 8.7 Average uplink delays for Cameras and Sensors decrease as the numerology is increased

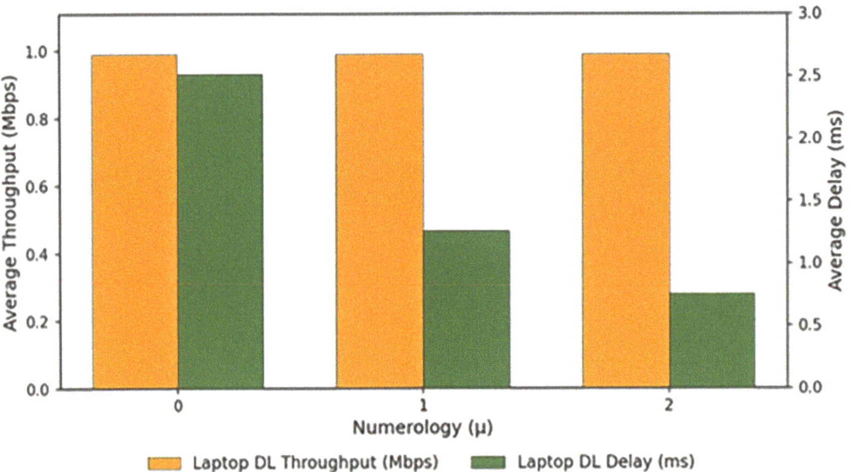

Fig. 8.8 Average downlink throughput for Laptops remains the same as the numerology is increased. This is because the flow is UDP. The average downlink delay decreases as the numerology is increased

at the UE in the same way across all numerologies, and the throughput becomes independent of μ. However, as μ increases, since each packet gets transmitted in a shorter time duration (because the packet occupies a single slot duration), the latency of communication reduces.

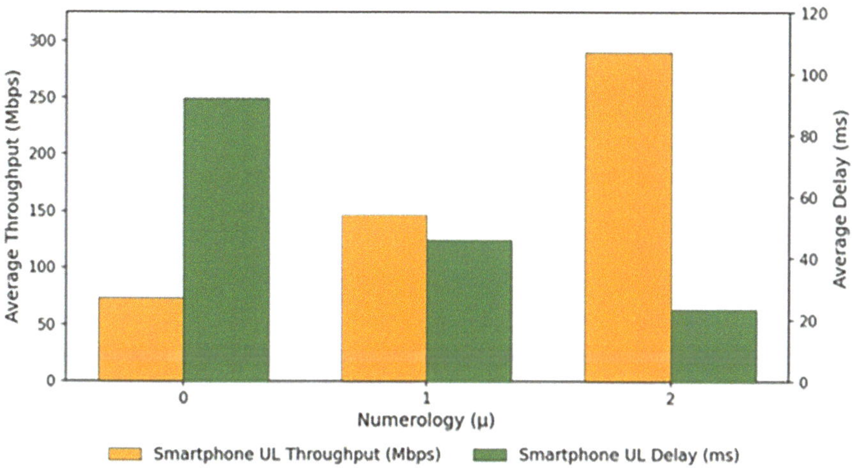

Fig. 8.9 Average uplink throughput for Smartphones increases as the numerology is increased. The average uplink delay decreases as the numerology is increased. This is because the flow is TCP

In contrast, in the case of TCP, the throughput crucially depends on the round-trip time involved in completing the three-way handshake mechanism of TCP. This is because the gNB sends the next packet only after it receives an acknowledgment from the UE for the successful transmission of the current packet. Now, if the current packet can be transmitted within one slot (across all values of μ), then the time taken to transmit the packet to the UE as well as the time taken to receive the feedback from the UE both reduce by a factor of 2 when we go from μ =0 to μ =1; as a consequence, the throughput roughly doubles as μ and is increased from 0 to 1. Similarly, the latency is reduced by a factor of 2, as μ is increased from 0 to 1, which can be explained similarly to that observed in the UDP scenario.

Therefore, the selection of the numerology in an NR system should take the traffic patterns into account.

8.6 Exercises

1. Observe how the results change when the propagation delays of the links are varied.

Appendix

We provide the implementation details of this experiment here. These are specific to NetSim v13.3.

Procedure to Import the Workspace for the Experiment

1. Follow the same steps in 1–8 of the Appendix in Chap. 4 and get to the stage as shown in Fig. 8.10.

Setting the Network Configurations and Executing the Two Cases

- *CASE—I*

1. The various properties (including the gNB and UE) can be configured in the physical layer of the 5G RAN properties of the gNB and UE items. The specific parameters are shown in Tables 8.7, 8.8, 8.9, and 8.10.
2. The Tx_Antenna_Count was set to 2 and Rx_Antenna_Count was set to 2 in gNB > Interface 5G_RAN > Physical Layer.
3. Run simulation for 10 s. After the simulation completes, go to the metrics window and note down the throughput and delay value from the application metrics.

- *CASE—2*

1. The various properties (including the gNB and UE) can be configured in the physical layer of the 5G RAN properties of the gNB and UE items. The specific parameters are shown in Tables 8.11, 8.12, 8.13, 8.14, 8.15, and 8.16.
2. The Tx_Antenna_Count was set to 2, and the Rx_Antenna_Count was set to 2 in gNB > Interface 5G_RAN > Physical Layer.

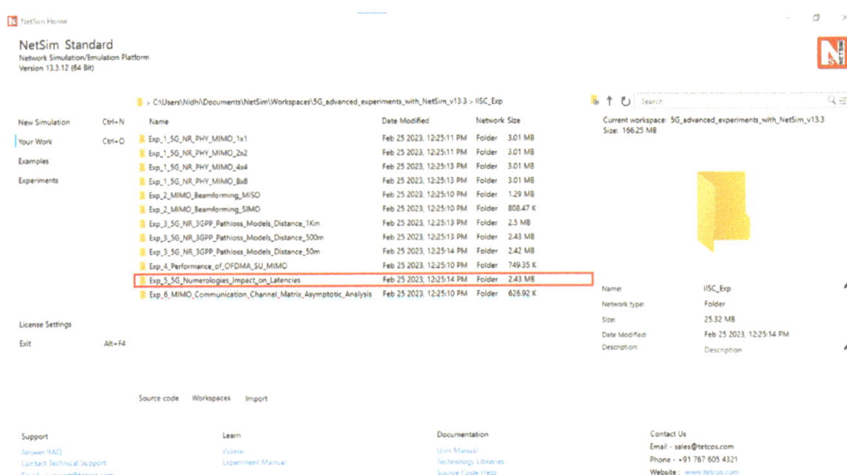

Fig. 8.10 NetSim Your Work Window with the experiment folders inside the workspace

Table 8.7 Physical layer properties set in the 5G RAN interface of gNB

gNB Properties - > Interface (5G_RAN)	
Pathloss model	None
Frequency range	FR1
CA type	Single band CA
CA configuration	n78
Numerology	0,1 and 2
Channel Bandwidth	10 MHz
DL: UL ratio	1:1
MCS table	QAM64
CQI table	Table 1

Table 8.8 Wired link properties set for case-1 of this experiment

Link properties (All wired links)	
Uplink/downlink speed (Mbps)	10,000
Uplink/downlink BER	0
Uplink/downlink propagation delay (μs)	0

Table 8.9 Custom application properties for UL UDP

Custom UL UDP	
Generation rate (Mbps)	5.8
Transport protocol	UDP
Application type	Custom
Packet size (Bytes)	1460
IAT distribution	Exponential
Inter arrival time (μs)	2000

Table 8.10 Custom application properties for DL UDP

Custom DL UDP	
Generation rate (Mbps)	5.8
Transport protocol	UDP
Application type	Custom
Packet size (Bytes)	1460
IAT distribution	Exponential
Inter arrival time (μs)	2000

3. Run the simulation for 110 s. After the simulation completes, go to the metrics window and note down the throughput and delay value from the application metrics.

Table 8.11 Physical layer properties set in 5G RAN

gNB Properties - > Interface (5G_RAN)	
Pathloss model	None
Frequency range	FR1
CA type	Inter band CA
CA configuration	CA_2DL_2UL_n40_n41
CA1	
Numerology	0, 1 and 2
Channel bandwidth	50 MHz
DL: UL ratio	1:4
CA2	
Numerology	0, 1 and 2
Channel bandwidth	50 MHz
DL: UL ratio	2:4
MCS table	QAM64
CQI table	Table 1

Table 8.12 Phone applications for UL TCP

Phone UL TCP	
Application type	FTP
Transport protocol	TCP
Start time (s)	$4.5 + 4(i - 1)$ where, $i = 1,2, \ldots .25$
Stop time (s)	105
File size (B)	1,500,000
Phone DL TCP	
Application type	FTP
Transport protocol	TCP
Start time (s)	$1.5 + 4(i - 1)$ where, $i = 1,2, \ldots .25$
Stop time (s)	105
Phone UL TCP	2,50,00,000

Table 8.13 Wired link properties set for case-2 of this experiment

Link properties (All wired links)	
Uplink/downlink speed (Mbps)	10,000
Uplink/downlink BER	0
Uplink/downlink propagation delay (μs)	5

Table 8.14 Sensor application properties for UL UDP

Sensor UL UDP	
Generation rate (Mbps)	1.6
Transport protocol	UDP
Application type	Custom
Packet size (Bytes)	500
Inter arrival time (μs)	8000

Table 8.15 Camera application properties for UL UDP

Camera UL UDP	
Generation rate (Mbps)	5
Transport protocol	UDP
Application type	Custom
Packet size (Bytes)	500
Inter arrival time (μs)	2000

Table 8.16 Laptop application properties for DL UDP

Laptop DL UDP	
Generation rate (Mbps)	2
Transport protocol	UDT
Application type	Custom
Packet size (Bytes)	1460
Inter arrival time (μs)	11680

Steps to Log the Simulation Results

The results will appear in the Application Metrics table. There is no need for any log files to be opened in this experiment.

References

Patriciello N, Lagen S, Giupponi L, Bojovic B (2018) 5G new radio numerologies and their impact on the end-to-end latency. In: IEEE 23rd international workshop on computer aided modeling and design of communication links and networks (CAMAD).

Zaidi AA et al (2016) Waveform and numerology to support 5G services and requirements. IEEE Commun Magaz 54(11):90–98

Chapter 9
Understanding the Role of Interference in 5G Network Design: Interference Modeling in Handover Procedures

9.1 Objective

In this experiment, we will study the impact of downlink interference on the signal-to-interference ratio (SINR) via NetSim v13.3. Particularly, we will study the following:

A. We consider a handover procedure in a cellular system and analyze the following cases:

1. The handover of a UE without interference, with pathloss exponent $\eta = 2.5$,
2. The handover of a UE without interference, with pathloss exponent $\eta = 4$,
3. The handover of a UE with interference, with pathloss exponent $\eta = 2.5$,
4. The handover of a UE with interference, with pathloss exponent $\eta = 4$.

B. To understand the effect of pathloss exponents with and without interferences on the point of handovers (locations where handovers occur) in cellular systems.

In this experiment, we consider the following handover scenario: A UE starts from BS_1 and moves in a straight line to BS_2. When the UE is attached to BS_1, it experiences interference from BS_2. Once it gets handed over to BS_2, the UE experiences interference from BS_1. There is always, thus, only one interferer, and we analyze the SINR as the UE moves a straight line from BS_1 to BS_2, as shown in Fig. 9.1.

9.2 Introduction

Due to the scarcity of the wireless spectrum, it is not possible for 5G networks to provide non-overlapping frequency bands for all communications that concurrently occur. Some transmissions will necessarily occur at the same time in the same frequency band, separated only in space, and the signals operating on the same time–frequency resources from many undesired or interfering transmitters are added to the

L. Yashvanth et al., *Understanding 5G New Radio*, Transactions on Computer Systems and Networks, https://doi.org/10.1007/978-981-92-0112-9_9

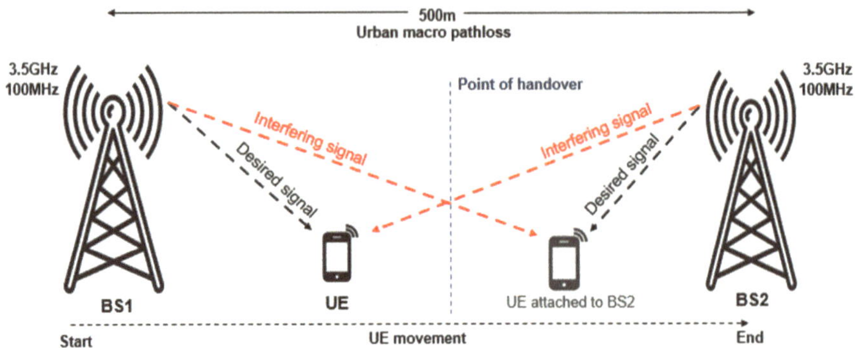

Fig. 9.1: UE1 is initially attached to BS1. The signal from BS1 is the desired signal, while the signal from BS2 is the interfering signal. Post-handover, the desired signal is transmitted from BS2, while that from BS1 is the interfering signal. We assume omnidirectional antennas at both BSs and consider cases with $\eta = 2.5, 4$

desired transmitter's signal at a receiver. The main determinants of the interference at a UE are.

- The network geometry, i.e., the location of the receivers and the transmitters,
- Base stations' transmit power,
- The pathloss model (how the signal attenuates with distance).

The performance and coverage of a 5G network critically depend on the signal-to-interference-and-noise ratios (SINRs) level at the receivers (Molisch 2012). This is defined as

$$\text{SINR} = \frac{P_r}{N_0 W + I},$$

where P_r is the received power of the desired signal, W is the bandwidth, N_0 is the thermal noise power spectral density, and I is the sum received power of all interfering signals. In 5G, the MCS is computed from the SINR. The higher the SINR, the higher the MCS, and hence the higher the data rate. Interference is, therefore, an important performance-limiting factor in wireless networks, and hence, it is crucial to characterize the effect of interference.

9.3 Network Scenario

Network Configurations and Sample Results for ISD = 500 m, C Band, 100 MHz, Urban Macro

In our network scenario, the inter-site distance (ISD) between BS_1 and BS_2 is 500 m. Both gNBs operate in the 3.5 GHz band, called the C band, with a bandwidth of

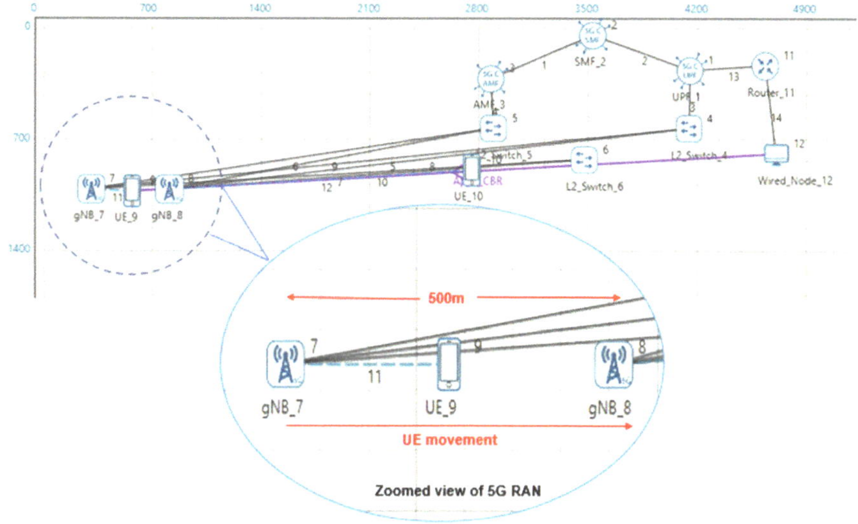

Fig. 9.2 NetSim scenario during mobility

100 MHz. The environment is assumed to be urban with signal attenuation as per the 3GPP Urban Macro pathloss model. Shadow-fading and fast fading are turned off to avoid sources of randomness.

Case#1: No interference in both base stations with $\eta = 2.5$

The network scenario is shown in Fig. 9.2 above.

9.4 Network Configuration

The distance between the two gNBs is 500 m, and a mobility profile is set accordingly. The coordinates of different devices are set as per Table 9.1.

1. The gNB properties are set as per Table 9.2.
2. The UE properties are set as per Table 9.3.

Table 9.1 Device coordinates

Devices	X	Y
gNB_8	800	200
gNB_7	300	200
UE_9	300	200
UE_10	300	760

Table 9.2 Values set for different parameters in simulation

gNB properties	
gNB height (m)	10 m
Tx power (dBm)	40 dBm
Tx antenna count	1
Rx antenna count	1
CA type	Single band
CA configuration	n78 (C band)
F_low (MHz)	3300
F_high (MHz)	3800
DL: UL ratio	4:1
Numerology	2
Channel bandwidth (MHz)	100MHZ
MCS table	QAM64
CQI table	TABLE1
Outdoor scenario	Urban Macro
Indoor office type	Mixed Office
PathLoss model	LOG_DISTANCE
PathLoss exponent	2.5
Shadow fading model	None
Fading and beamforming	NO FADING MIMO UNIT GAIN
Downlink interference model	No interference

Table 9.3 UE properties

UE properties	
UE height	1.50
Tx power	23 dBm
Tx antenna count	1
Rx antenna count	1

9.5 Sample Results

The values of SINR seen by the UE from gNB_7 and gNB_8 are given in Tables 9.4 and 9.5, respectively.

Case #2: No interference in both base stations with $\eta = 4$.

In this case, the pathloss of the environment is increased to 4 from 2.5 in the previous case. We provide the results for this case in a consolidated manner with other cases.

Case #3: UE to both base stations is in $\eta = 2.5$ (interference present).

Table 9.4 Downlink SINR values from GNB_7, with ISD = 500 m

Distance	Max of SINR (dB)
0	62.43
10	62.43
20	56.95
30	53.03
40	50.09
50	47.75
60	45.82
70	44.17
80	42.74
90	41.48
100	40.34
110	39.31
120	38.37
130	37.51
140	36.71
150	35.96
160	35.26
170	34.61
180	33.99
190	33.40
200	32.85
210	32.32
220	31.81
230	31.33
240	30.87
250	30.43
260	30.00
270	29.59
280	29.20
290	28.82

In this case, while a UE is being served by one BS, the interference from the neighboring BS is also made active such that $\eta = 2.5$ is set in all the links. This, in turn, may potentially affect the SINR and, hence, the point of handovers. We discuss all these effects below after presenting the results.

Case #4: UE to both base stations is in $\eta = 4$ ***(interference present).***

This case is the same as Case 3, except that η is set to 4.

Table 9.5 Downlink SINR from GNB_8, with ISD = 500 m

Distance	Max of SINR (dB)
290	32.32
300	32.85
310	33.40
320	33.99
330	34.61
340	35.26
350	35.96
360	36.71
370	37.51
380	38.37
390	39.31
400	40.34
410	41.48
420	42.74
430	44.17
440	45.82
450	47.75
460	50.09
470	53.03
480	56.95
490	62.43
500	62.43

The results for cases 2, 3, and 4 are reported in Table 9.6, and a pictorial illustration of the measured SINR at the UE for all the four cases is given in Fig. 9.3.

9.6 Discussion

Initially (in case 1), the UE is attached to BS_1. Now, when the UE moves in a straight line toward BS_2, at 290 m, the UE gets handed over to BS_2. Till 290 m, the "desired" signal is from BS_1 while the "interfering" signal is from BS_2. Once the handover happens, the scenario reverses due to the handover, and the desired signal arrives from BS_2 while the interfering signal is caused by BS_1. Now, to understand the trend in the experienced SINR in Fig. 9.3, we first explain the method by which the SINR is computed in this experiment.

First of all, recall that the signals from BS_1 and BS_2 that arrive at the UE get attenuated by the pathloss that depends on the distance between the UE and the BSs. Specifically, if the transmit powers at both BS s are P_t, then the $SINR$ at the UE can

Table 9.6 Results for SINR versus distance for cases 2, 3, and 4 for ISD-500 m downlink

Distance	Case #2 DL SINR (dB)	Case #3 DL SINR (dB)	Case #4 DL SINR (dB)	Distance	Case #2 DL SINR (dB)	Case #3 DL SINR (dB)	Case #4 DL SINR (dB)
0	45.66	39.50	45.58	290	– 2.52	**3.49**	– 3.14
10	45.66	39.28	45.58	300	– 1.68	4.39	– 2.22
20	36.90	33.59	36.81	310	– 0.79	5.30	– 1.27
30	30.63	29.44	30.53	320	0.15	6.23	– 0.28
40	25.92	26.26	25.81	330	1.14	7.18	0.76
50	22.18	23.69	22.06	340	2.19	8.16	1.85
60	19.08	21.51	18.96	350	3.31	9.18	3.01
70	16.45	19.62	16.31	360	4.50	10.23	4.23
80	14.16	17.93	14.01	370	5.79	11.33	5.54
90	12.13	16.40	11.97	380	7.17	12.48	6.95
100	10.32	15.00	10.14	390	8.67	13.70	8.47
110	8.67	13.70	8.47	400	10.32	15.00	10.14
120	7.17	12.48	6.95	410	12.13	16.40	11.97
130	5.79	11.33	5.54	420	14.16	17.93	14.01
140	4.50	10.23	4.23	430	16.45	19.62	16.31
150	3.31	9.18	3.01	440	19.08	21.51	18.96
160	2.19	8.16	1.85	450	22.18	23.69	22.06
170	1.14	7.18	0.76	460	25.92	26.26	25.81
180	0.15	6.23	– 0.28	470	30.63	29.44	30.53
190	– 0.79	5.30	– 1.27	480	36.90	33.59	36.81
200	– 1.68	4.39	– 2.22	490	45.66	39.28	45.58
210	– 2.52	3.49	– 3.14	500	45.66	39.50	45.58
220	– 3.33	2.61	– 4.04	–	–	–	–
230	– 4.10	1.73	– 4.91	–	–	–	–
240	– 4.84	0.86	– 5.77	–	–	–	–
250	– 5.55	0.00	– 6.61	–	–	–	–
260	– 6.23	– 0.87	– 7.46	–	–	–	–
270	– 6.88	– 1.74	– 8.31	–	–	–	–
280	– **7.51**	– 2.62	– **9.17**	–	–	–	–
280	– **3.33**	– 2.62	– **4.04**	–	–	–	–
290	– 2.52	– **3.50**	– 3.14	–	–	–	–

The bold entries indicate the points of handover

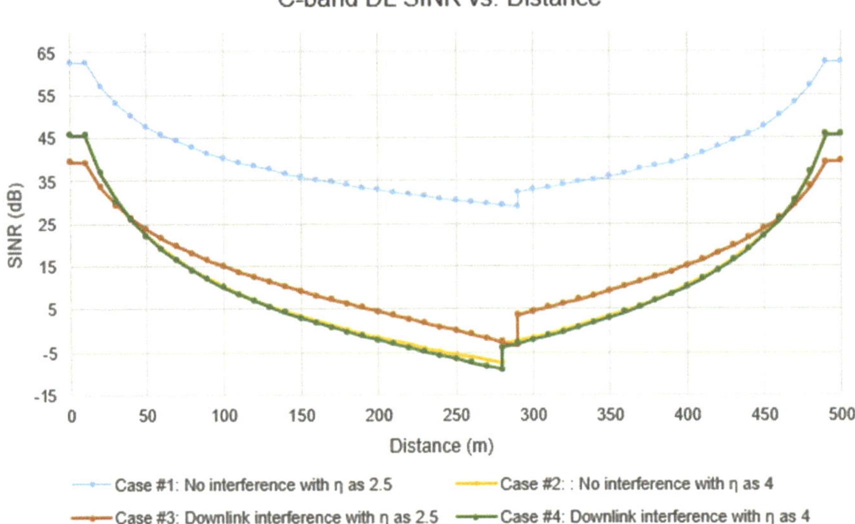

Fig. 9.3 Downlink SINR versus distance plot for C-band for different network configurations

be evaluated as:

$$SINR = \frac{P_t \times PL(d)}{N_0 \times W + P_t \times PL(d_{ISD} - d)},$$

where $PL(d)$ is the pathloss loss (modeled as per the 3GPP pathloss models) at a distance of d from BS_1. Since the distance between the two BSs, (also called the inter-site distance) is equal to d_{ISD}, the UE is at a distance of $(d_{ISD} - d)$ from the interfering BS; thus, we compute the pathloss from interfering BS as the $PL(d_{ISD} - d)$ term in the denominator. When there is a line-of-sight (LOS) link with a gNB, the pathloss is lower, which is modeled in this experiment by setting the pathloss exponent to $\eta = 2.5$. When there is a non-line-of-sight (NLOS) link with a gNB, the pathloss is higher, which is modeled by setting the pathloss exponent to $\eta = 4$.

With this background, we make the following observations from Fig. 9.3:

- In all cases, we see a constant SINR till 10m because the pathloss equations defined in the standard take effect only from 10m, and until then, we do not account for pathloss.
- Now note that we plot the SINR versus distance for the following four cases:
  - Case #1: UE is in LOS with both BSs, and interference is turned off,
  - Case #2: UE is in NLOS with both BSs, and interference is turned off,
  - Case #3: UE is in LOS with both BSs, with interference turned on,
  - Case #4: UE is in NLOS with both BSs, with interference turned on.

- In case 1 and case 2, the term I in $SINR = \frac{P_r}{N_0 W + I}$ is set to zero. Operationally, this indicates that the two BSs operate in non-overlapping frequency bands, and therefore, we have $SINR = SNR = \frac{P_r}{N_0 W}$. Now, in the figure, we see the SNR dropping as the UE moves away from BS_1. At 280/290m, it gets handed over to BS_2, and we see the SNR increasing as the UE moves closer to BS_2. Notably, we observe an abrupt "jump" at the handover point. This is because the 3GPP standards specify that handover should occur only when the target-gNB's SINR is higher than the serving-gNB's SINR by a pre-specified offset (e.g., 3 dB), called the *handover margin*. The handover margin ensures that the UE does not undergo a "ping-pong" handover arising due to local fluctuation of the SINR. In particular, by specifying a handover margin, the UE gets sufficient time to average out the received SINR, which can then be used to reliably decide whether a handover is needed in the first place (Dutta and Schulzrinne 2014). More details on this are provided in Chap. 10.
- Next, we observe that the NLOS curve (yellow) is lower than the LOS curve (blue). This is because NLOS pathloss is higher than the LOS pathloss.
- In cases 3 and 4, the observations are similar to the above, but the SINR values are lower than their respective counterparts in cases 1 and 2, due to additional interferences that degrade the SINR further. Further, we notice that the two curves of cases 2 and 4 are very close to each other. This is because the decay in the SINR due to higher pathloss exponents outweighs the decay of SINR due to the interference power, which explains the almost similar performance. In other words, in NLOS, while the signal power decays more rapidly with distance, so does the interference power, making the effect of interference much lower.
- As a final observation, we note that the handover points are different when the pathloss exponents change, and we explain this in detail next.

Understanding the Points of Handoff

For the sake of exposition, we investigate the point of handoff for case #1 and case #2 where the interferences are assumed to be absent from the base stations. This, therefore, represents a noise-limited regime, or a scenario where the two BSs use non-overlapping frequency bands. In such a scenario, as the UE moves from BS_1 toward BS_2, the SNR from BS_1 decreases, while the SNR from BS_2 increases. The point where the SNR from BS_2 is 3 dB higher than that from BS_1 determines the point of handoff. But we observe that between the cases with pathloss exponents of 2.5 and 4, the points of handovers are different! To understand this, for the sake of generality, we discuss the effect of the pathloss exponent on handover in a general setting, independent of the values configured/obtained in this experiment. Now, consider the following noise-limited scenario:

Let a UE move from base station A toward base station B with inter-site distance = 1 km as shown in Fig. 9.4. Let the pathloss exponent be set to 1 ($\eta = 1$), then the solid blue curve represents the received SNR from BS A, while the solid purple curve represents the received SNR from BS B. Further, assume that Δ_{HO} is the "handover margin", i.e., the required SNR difference between the base stations to perform a

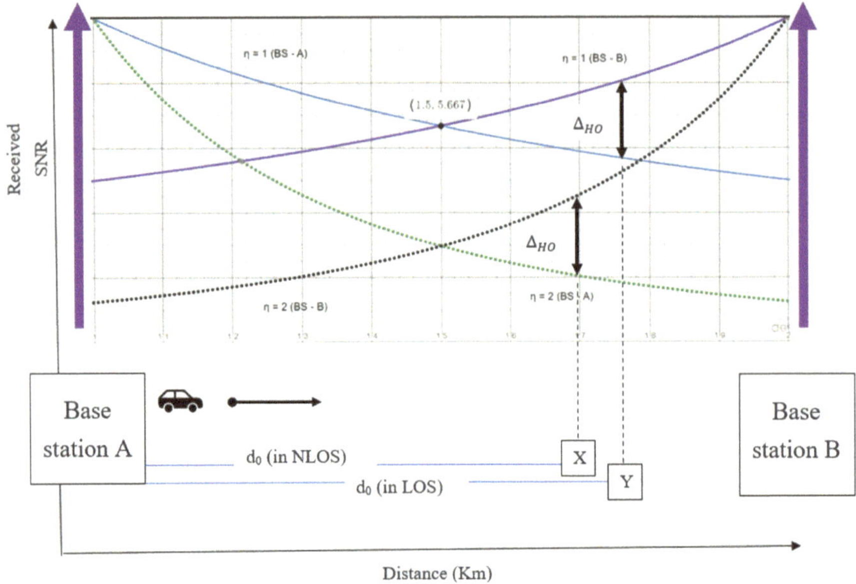

Fig. 9.4 Illustration of different handover points in LOS/NLOS cases

handoff from BS A to BS B. Let the distance at this point be d_0. This point is indicated by point Y in the above plot.

Now, when the pathloss exponent increases to 2, i.e., $\eta = 2$, the received SNR from both the BSs changes, and the corresponding curves are shown in the above figure using dashed lines. Due to the differing slopes in the SNR curves at different values of η while the handover margin remaining the same, the point at which the handoff occurs indeed becomes different (in fact, it occurs earlier than the former case as shown in the figure), and this is indicated by point X. In fact, observe from the figure that, as we increase the pathloss from 1 to 2, the point of handoff changes from d = 1.7 km to d = 1.76 km, respectively. This difference in handover points is more pronounced when the difference in the pathlosses increases. Similar observations can be drawn in interference-limited scenarios also.

Conclusion: The point of handoff is different for environments with different pathloss exponents for a given handoff threshold.

9.7 Exercises

1. Set a different inter-gNB distance and an appropriate mobility profile, and study the handover mechanism as discussed above.
2. Repeat this experiment for a different frequency band (say n5 or low band—e.g., 850 MHz). Justify your change in the observations.

3. Enable additional channel fluctuations (by turning the shadowing on in NetSim), and observe the ping-pong effects during the handover phase.

Appendix

We provide the implementation details of this experiment here. These are specific to NetSim v13.3.

Setting up the Experiment Configuration Files

- ***Case #1: No Interference with*** $\eta = 2.5$

Procedure:

1. Open NetSim v13.3, available on the desktop.
2. The first task is to create a new workspace.
3. On the "**Your Work**" window, select "**Workspaces,**" and click on "**List of Workspaces,**" as shown in Fig. 9.5.
4. Then click on "**New Workspace,**" name your workspace, and **locate it in your own folder.** See Fig. 9.6.
5. Use the download link given in the **Appendix-A** to download the netsim workspace file (*.netsimexp). for this experiment.
6. Go to the NetSim Home window, go to Your Work, and click on Import.
7. In the Import Workspace Window, browse and select the *.netsimexp file from the extracted directory. See Fig. 9.7.
8. The configuration files for the four scenarios of this experiment will appear on the screen, click on the first case and perform the experiment.

Setting the Network Configuration and Executing the Experiment

1. Place the two gNBs in the grid environment and set the distance between them to 500m.
2. Go to gNB and UE properties. In the RAN interface, set the physical layer properties of the gNB and the two UEs as given in Tables 9.2 and 9.3.
3. In the General layer, set the UE's X and Y coordinates as the gNB1's X and Y coordinates. That is, the initial position of UE is the same as the location of gNB1.
4. Set the mobility model as file-based mobility and configure the mobility that UE needs to travel straight toward the gNB2. So, configure the mobility file according to the distance that UE needs to travel. For example, in the above scenario, UE needs to travel 500m from gNB1 to gNB2. Give this as an input in the Excel sheet as shown in Fig. 9.8.

1. Before clicking on Run, select the "logs" option and check "LTENR Radio Measurements Log." See Fig. 9.9.

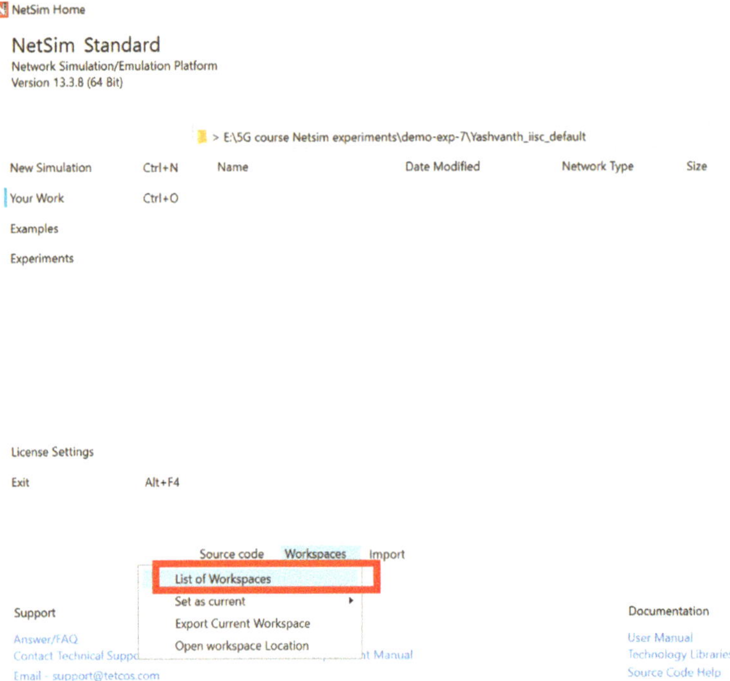

Fig. 9.5 "Your Work" Page of NetSim

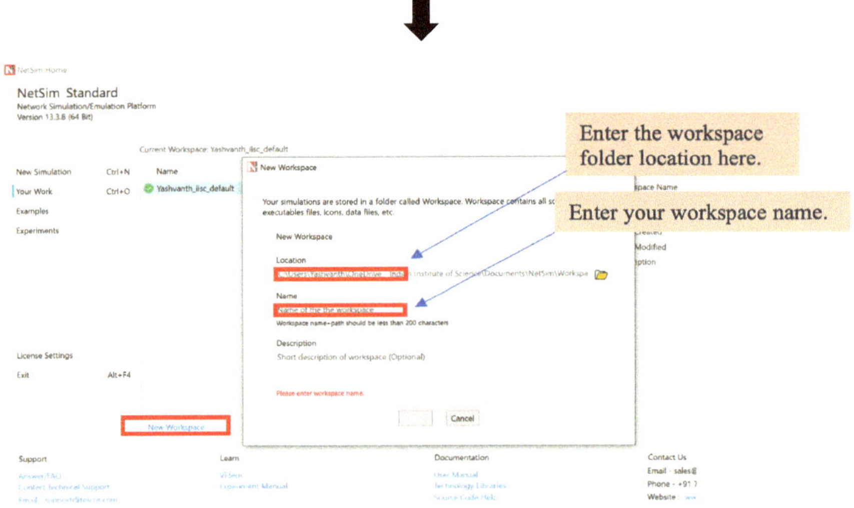

Fig. 9.6 Creating a new Workspace

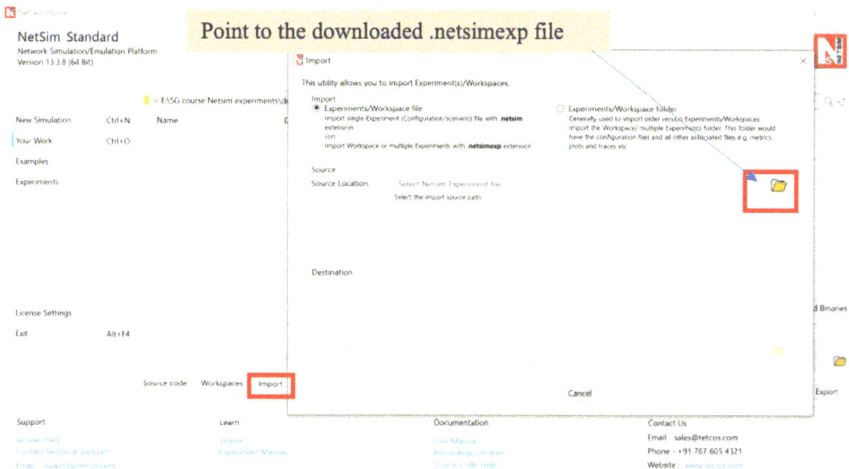

Fig. 9.7 Loading the workspace

	A	B	C	D	E
1	#Time(s)	Device ID	X	Y	Z
2	1	9	310	200	0
3	1.2	9	320	200	0
4	1.4	9	330	200	0
5	1.6	9	340	200	0
6	1.8	9	350	200	0
7	2	9	360	200	0
8	2.2	9	370	200	0
9	2.4	9	380	200	0
10	2.6	9	390	200	0
11	2.8	9	400	200	0
12	3	9	410	200	0
13	3.2	9	420	200	0
14	3.4	9	430	200	0
15	3.6	9	440	200	0
16	3.8	9	450	200	0
17	4	9	460	200	0
18	4.2	9	470	200	0
19	4.4	9	480	200	0
20	4.6	9	490	200	0
21	4.8	9	500	200	0
22	5	9	510	200	0
23	5.2	9	520	200	0
24	5.4	9	530	200	0
25	5.6	9	540	200	0
26	5.8	9	550	200	0
27	6	9	560	200	0

	A	B	C	D	E
25	5.6	9	540	200	0
26	5.8	9	550	200	0
27	6	9	560	200	0
28	6.2	9	570	200	0
29	6.4	9	580	200	0
30	6.6	9	590	200	0
31	6.8	9	600	200	0
32	7	9	610	200	0
33	7.2	9	620	200	0
34	7.4	9	630	200	0
35	7.6	9	640	200	0
36	7.8	9	650	200	0
37	8	9	660	200	0
38	8.2	9	670	200	0
39	8.4	9	680	200	0
40	8.6	9	690	200	0
41	8.8	9	700	200	0
42	9	9	710	200	0
43	9.2	9	720	200	0
44	9.4	9	730	200	0
45	9.6	9	740	200	0
46	9.8	9	750	200	0
47	10	9	760	200	0
48	10.2	9	770	200	0
49	10.4	9	780	200	0
50	10.6	9	790	200	0
51	10.8	9	800	200	0

Fig. 9.8 UE mobility file for 500 m

2. Now, Run the Simulation for 12 s.

Steps to Log the Simulation Results

1. After simulation, open the LTENR Radio measurement log present under the Log files section in the Results Dashboard, as shown in Fig. 9.10.
2. Create a pivot table for this log file by clicking the pivot option at the top of the ribbon under the Insert section, as shown in Fig. 9.11.
3. In the pivot table drop 'gNB/eNB Name', 'UE Name', 'Channel' fields under filters area, drop 'Distance' in Row area and drop 'SINR' in Value area. Set the SINR values to max by clicking on the arrow icon present at the end of the field - > value field setting- > max, as shown in Fig. 9.12.

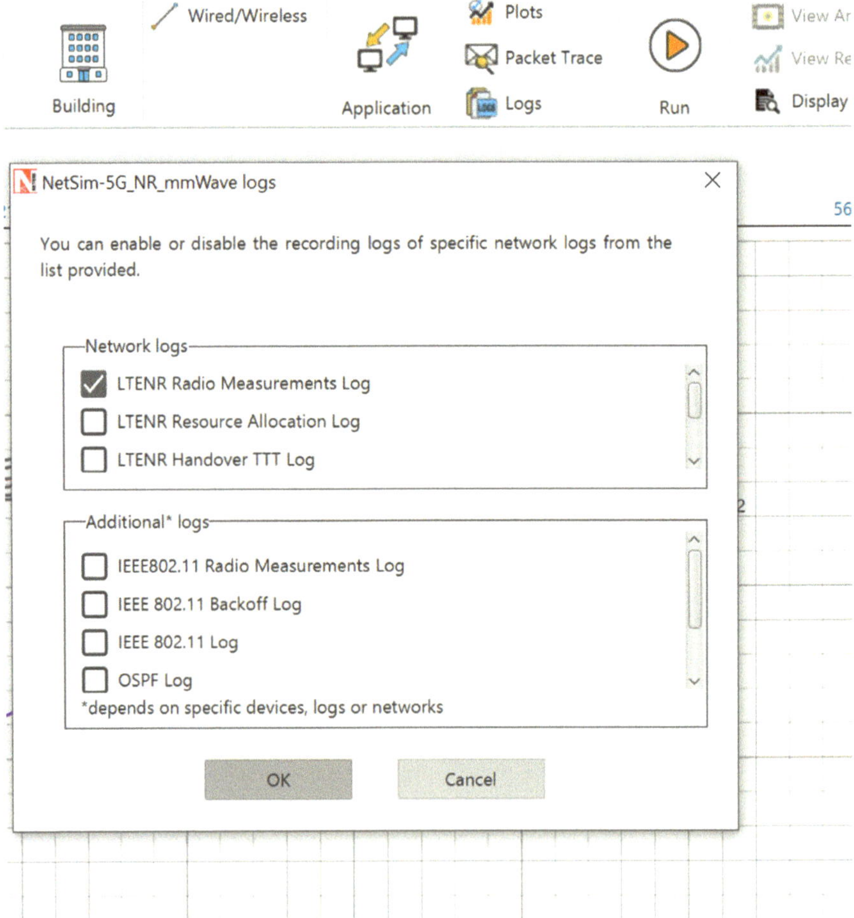

Fig. 9.9 Enabling the Log options

Fig. 9.10 ISD 500 m simulation result dashboard

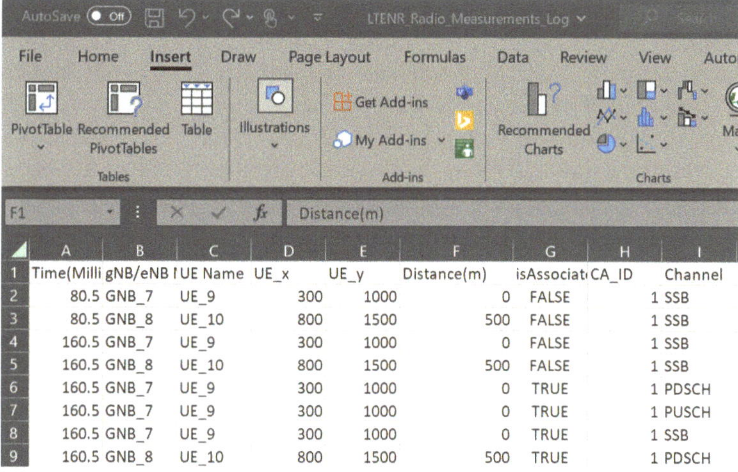

	A	B	C	D	E	F	G	H	I
1	Time(Milli	gNB/eNB I	UE Name	UE_x	UE_y	Distance(m)	isAssociat	CA_ID	Channel
2	80.5	GNB_7	UE_9	300	1000	0	FALSE	1	SSB
3	80.5	GNB_8	UE_10	800	1500	500	FALSE	1	SSB
4	160.5	GNB_7	UE_9	300	1000	0	FALSE	1	SSB
5	160.5	GNB_8	UE_10	800	1500	500	FALSE	1	SSB
6	160.5	GNB_7	UE_9	300	1000	0	TRUE	1	PDSCH
7	160.5	GNB_7	UE_9	300	1000	0	TRUE	1	PUSCH
8	160.5	GNB_7	UE_9	300	1000	0	TRUE	1	SSB
9	160.5	GNB_8	UE_10	800	1500	500	TRUE	1	PDSCH

Fig. 9.11 Inserting pivot table

4. Now filter the gNB as gNB_7 (from gNB_8), UE name as UE_9, and Channel as PDSCH, as shown in Fig. 9.13.
5. Copy the values from 0 to 290 along with the Row Labels and Max_SINR_dB header and paste them into another sheet. This will complete obtaining the results as shown in Table 9.4. Similarly, filter gNB/eNB name to gNB_8 and copy the row label value along with SINR, and paste it next to the previously created new sheet.
6. Next, we note that NetSim always calculates the distance of a UE from its attached gNB. In the plot that we eventually wish to obtain, the X-axis has a distance from

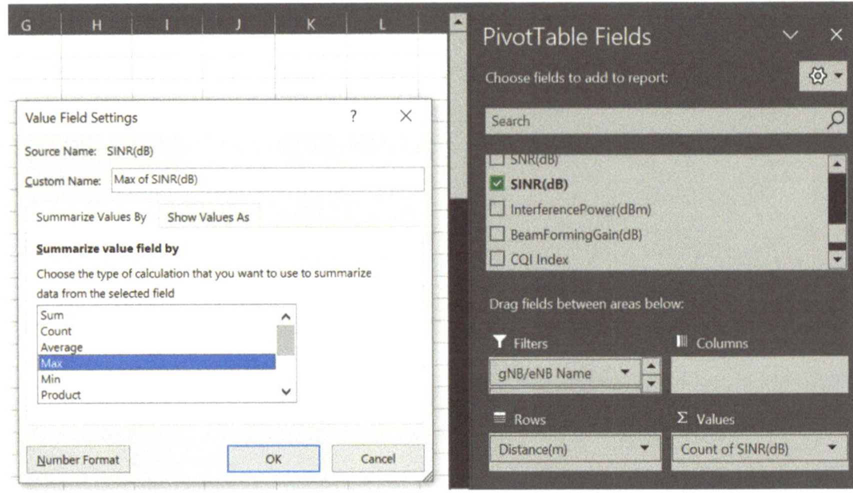

Fig. 9.12 Creating pivot table

the initially attached gNB, which is gNB_7. In our experiment, UE_9 is initially attached to gNB_7, and post-handover, it gets attached to gNB_8. Since the gNB_7 to gNB_8 distance is 500m, post-handover, the distance of UE_9 from the initial gNB_7 is $500 - d^{UE(9)}_{gNB(8)}$, i.e., $d^{UE(9)}_{gNB(7)} = 500 - d^{UE(9)}_{gNB(8)}$. Thus, after the handover point, ensure to copy the row values from 0 to 280 under the GNB_8 readings. This will complete obtaining the results as shown in Table 9.5.

7. Finally, appending the measurements obtained in the previous two steps and arranging them in the distance of 0 to 500 m of the UE_9 measured from the GNB_7 will help with plotting the curve shown in Fig. 9.3.
8. By following the above steps, we can similarly obtain the results for cases 2, 3, and 4.

AutoSave Off LTENR_Radio_Measurements_Log.csv

File Home Insert Page Layout Formulas Data Review View

D1 f_x =500-E1

	A	B	C	D	E	F	G
1	gNB/eNB Name	GNB_8		500	0	62.42933	
2	UE Name	UE_9		490	10	62.42933	
3	Channel	PDSCH		480	20	56.95417	
4				470	30	53.0341	
5	Max of SINR(dB)			460	40	50.09006	
6	Distance(m)	Total		450	50	47.75241	
7	0	62.42933		440	60	45.81967	
8	10	62.42933		430	70	44.17441	
9	20	56.95417		420	80	42.74313	
10	30	53.034096		410	90	41.47705	
11	40	50.090055		400	100	40.34224	
12	50	47.752405		390	110	39.31418	
13	60	45.819667		380	120	38.37462	
14	70	44.174408		370	130	37.50958	
15	80	42.743129		360	140	36.70815	
16	90	41.477049		350	150	35.96163	
17	100	40.342239		340	160	35.26302	
18	110	39.314184		330	170	34.60654	
19	120	38.374619		320	180	33.98742	
20	130	37.509577		310	190	33.40163	
21	140	36.708145		300	200	32.84577	
22	150	35.961634		290	210	32.31695	
23	160	35.263021					
24	[illegible]	[illegible]					

Fig. 9.13 Copying the distance and SINR values into new cells

References

Dutta A, Schulzrinne H (2014) Mobility protocols and handover optimization: design. Wiley, Evaluation and Application

Molisch AF (2012) Wireless communications, vol 34. John Wiley & Sons

Chapter 10
Understanding Handover Mechanisms in 5G NR

10.1 Objective

In this experiment, we will study the procedure of handovers in 5G networks in more detail compared to Chap. 9. In particular, we will study the following aspects by performing simulation experiments using NetSim v13.3:

1. The process of handover signaling via packet analysis through the RAN and core-network.
2. The impact of handover on the delay and throughput of the UE under handover.

10.2 Introduction

The handover mechanism is generally based on the strongest adjacent cell handover algorithm (Dimou et al. 2009), which is also used by the NetSim 5G library. The algorithm enables each UE to connect to the gNB that provides the highest reference signal received power (RSRP). Therefore, a handover occurs when a gNB (possibly in an adjacent cell) offering a stronger RSRP (measured as SNR in NetSim) is detected.

NetSim implements a handover procedure similar to that described in 3GPP TS 38.331, Sec. 5.5.4.4, Event A3, wherein a handover occurs when a neighbor cell's RSRP becomes offset better than the serving cell's RSRP. Note that in NetSim, the report-type is periodic, not event-triggered, since NetSim is a discrete event simulator, not a continuous time simulator.

This algorithm is susceptible to ping-pong handovers, continuous handovers between the serving and adjacent cells on account of changes in RSRP due to mobility and shadow-fading (Dimou et al. 2009; Dutta and Schulzrinne 2014). At one instant, the adjacent cell's RSRP could be higher, and the very next, it could be the original serving cell's RSRP, and this continues.

To solve this problem, the handover algorithm uses:

L. Yashvanth et al., *Understanding 5G New Radio*, Transactions on Computer Systems and Networks, https://doi.org/10.1007/978-981-92-0112-9_10

(a) Hysteresis (Hand-over-margin, HOM) which adds a RSRP threshold (Adjacent_cell_RSRP—Serving_cell_RSRP > Hand-over-margin, or hysteresis), and
(b) Time-to-trigger (TTT), which adds a time-threshold.

This HOM is part of NetSim implementation while TTT can be implemented as a custom project in NetSim. The reader must refer to Chap. 9 for a discussion on the process of handovers and related theory from the PHY viewpoint.

10.3 Part I: Handover Algorithm

10.3.1 Network Scenario

The NetSim UI displays the network configuration for this experiment, shown in Fig. 10.1.

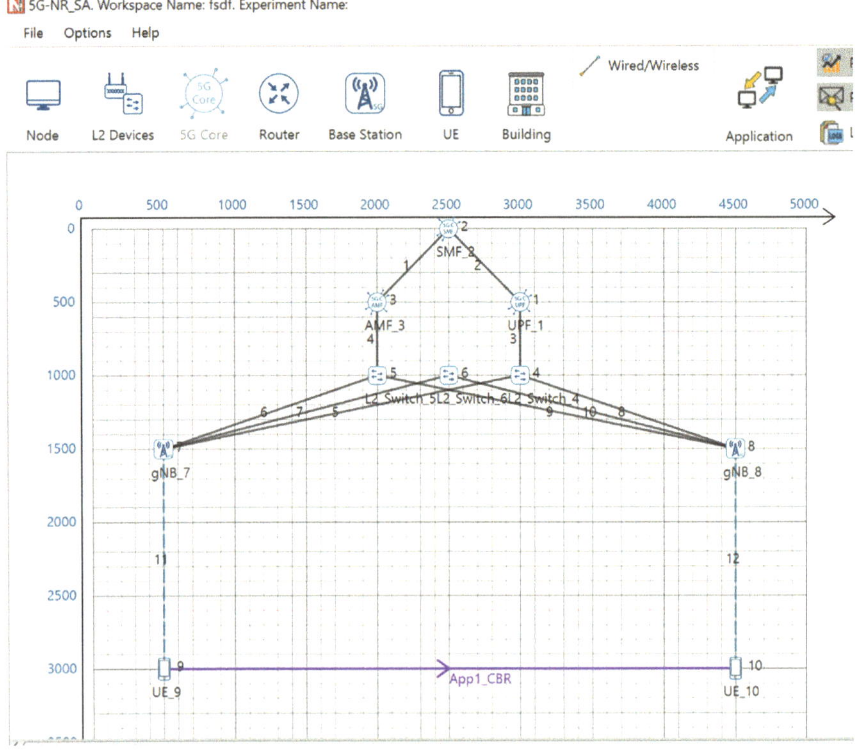

Fig. 10.1 Network set up for studying the 5G NR handover

10.3.2 Network Configuration

1. The device positions are set as per Table 10.1.
2. The properties of gNB 7 are chosen as per Table 10.2.
3. Similarly, the properties of gNB 8 are also set according to Table 10.2.
4. The transmit antenna counts are set to 2 at the gNBs, and the receive antenna counts to 1.

Table 10.1 Device positions for Part I

	gNB 7	gNB 8	UE 9	UE 10
X co-ordinate	500	4500	500	4500
Y co-ordinate	1500	1500	3000	3000

Table 10.2 gNB_7 properties

Interface_4 (5G_RAN) properties	
CA_Type	Single band
CA_Configuration	n78
CA_Count	1
Numerology	0
Channel bandwidth (MHz)	10
PRB count	52
MCS table	QAM64
CQI table	Table 1
X_Overhead	XOH0
DL UL ratio	4:1
Pathloss model	3GPPTR38.901-7.4.1
Outdoor scenario	Urban macro
LOS_NLOS_Selection	User_Defined
LOS probability	1
Shadow fading model	None
Fading_and_Beamforming	NO_FADING_MIMO_UNIT_GAIN
O2I building penetration model	LOW_LOSS_MODEL
Additional loss model	None

Table 10.3 Device positions for Part II

	gNB 7	gNB 8	UE 9	UE 10
X co-ordinate	500	4500	500	4500
Y co-ordinate	500	500	1000	1000

5. The transmit antenna counts are set to 2 at the UEs, and the receive antenna counts to 1.

10.3.3 Sample Results and Discussion

10.3.4 Handover Signaling

We depict the packet flow during the handover process in 5G NR in Fig. 10.2, and this can be observed from the packet trace of 5G NR, and a screenshot of it is shown in Fig. 10.3.

We now make the following explanations on the control packet flow between different nodes that happens during the handover process in 5G NR.

1. UE will send the UE_MEASUREMENT_REPORT every 120 ms to the connected gNB.
2. The initial UE-gNB connection and UE association with the core network takes place by transferring the radio resource control (RRC) and Registration, session request, and session response packets.
3. As per the configured file-based mobility, UE_9 moves toward gNB_8.
4. After 18.6 s, gNB_7 sends a HANDOVER REQUEST to gNB_8.
5. gNB_8 sends a HANDOVER REQUEST ACK to gNB_7.
6. After receiving the HANDOVER REQUEST ACK from gNB_8, gNB_7 sends the HANDOVER COMMAND to UE_9.
7. After the HANDOVER COMMAND packet is transferred to the UE, the target gNB will send a PATH SWITCH packet to the AMF via Switch 5.

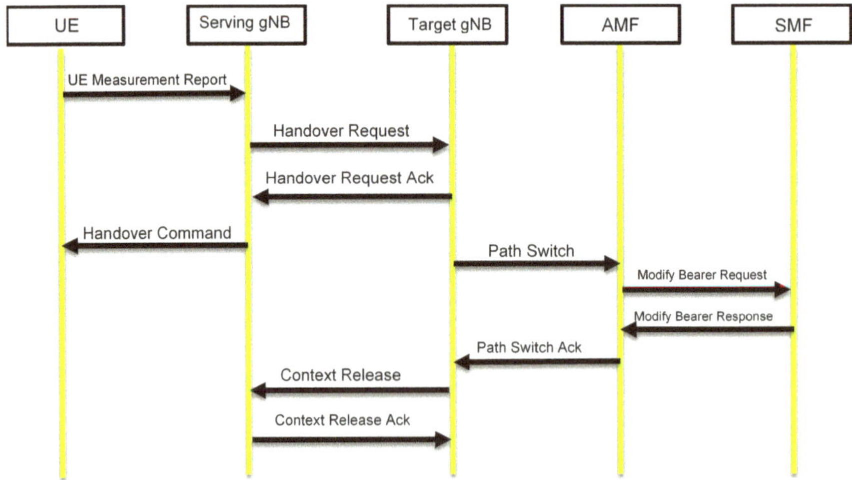

Fig. 10.2 Control packet flow in the 5G NR handover process

	CONTROL_PACKET_TYPE/APP_NAME	SOURCE_ID	DESTINATION_ID	TRANSMITTER_ID	RECEIVER_ID	APP_LAYER_ARRIVAL_TIME(µS)	TRX_LAYER_ARRIVAL_TIME(µS)	NW_LAYER_ARRIVAL_TIME(µS)	MAC_LAYE
3429	RRC_MIB	GNB-7	Broadcast-0	GNB-7	UE-10	N/A	N/A	N/A	
3430	RRC_SIB1	GNB-8	Broadcast-0	GNB-8	UE-9	N/A	N/A	N/A	
3431	RRC_SIB1	GNB-8	Broadcast-0	GNB-8	UE-10	N/A	N/A	N/A	
3432	RRC_MIB	GNB-8	Broadcast-0	GNB-8	UE-9	N/A	N/A	N/A	
3433	RRC_MIB	GNB-8	Broadcast-0	GNB-8	UE-10	N/A	N/A	N/A	
3434	UE_MEASUREMENT_REPORT	UE-9	GNB-7	UE-9	GNB-7	N/A	N/A	N/A	
3435	UE_MEASUREMENT_REPORT	UE-10	GNB-8	UE-10	GNB-8	N/A	N/A	N/A	
3436	UE_MEASUREMENT_REPORT	UE-9	GNB-7	UE-9	GNB-7	N/A	N/A	N/A	
3437	UE_MEASUREMENT_REPORT	UE-10	GNB-8	UE-10	GNB-8	N/A	N/A	N/A	
3438	HANDOVER_REQUEST	GNB-7	GNB-8	GNB-7	SWITCH-6	N/A	N/A	18600999	
3439	HANDOVER_REQUEST	GNB-7	GNB-8	SWITCH-6	GNB-8	N/A	N/A	18600999	
3440	HANDOVER_REQUEST_ACK	GNB-8	GNB-7	GNB-8	SWITCH-6	N/A	N/A	18601030.44	
3441	HANDOVER_REQUEST_ACK	GNB-8	GNB-7	SWITCH-6	GNB-7	N/A	N/A	18601030.44	
3442	HANDOVER_COMMAND	GNB-7	UE-9	GNB-7	UE-9	N/A	N/A	N/A	
3443	HANDOVER_COMMAND	GNB-7	UE-9	GNB-7	UE-9	N/A	N/A	N/A	
3444	PATH_SWITCH	GNB-8	AMF-3	GNB-8	SWITCH-5	18602999	18602999	18602999	
3445	PATH_SWITCH	GNB-8	AMF-3	SWITCH-5	AMF-3	18602999	18602999	18602999	
3446	MODIFY_BEARER_REQUEST	AMF-3	SMF-2	AMF-3	SMF-2	18603044.52	18603044.52	18603044.52	
3447	MODIFY_BEARER_RESPONSE	SMF-2	AMF-3	SMF-2	AMF-3	18603065.2	18603065.2	18603065.2	
3448	PATH_SWICTH_ACK	AMF-3	GNB-8	AMF-3	SWITCH-5	18603085.88	18603085.88	18603085.88	
3449	PATH_SWICTH_ACK	AMF-3	GNB-8	SWITCH-5	GNB-8	18603085.88	18603085.88	18603085.88	
3450	UE_CONTEXT_RELEASE	GNB-8	GNB-7	GNB-8	SWITCH-6	N/A	N/A	18603131.4	
3451	UE_CONTEXT_RELEASE	GNB-8	GNB-7	SWITCH-6	GNB-7	N/A	N/A	18603131.4	
3452	UE_CONTEXT_RELEASE_ACK	GNB-7	GNB-8	GNB-7	SWITCH-6	N/A	N/A	18603162.84	
3453	UE_CONTEXT_RELEASE_ACK	GNB-7	GNB-8	SWITCH-6	GNB-8	N/A	N/A	18603162.84	
3454	RRC_RECONFIGURATION	GNB-8	UE-9	GNB-8	UE-9	N/A	N/A	N/A	
3455	RRC_RECONFIGURATION	GNB-8	UE-9	GNB-8	UE-9	N/A	N/A	N/A	

Fig. 10.3 Screenshot of NetSim packet trace file showing the control packets involved in handover. *Note* Some columns have been hidden for ease of visualization

8. When the AMF receives the PATH SWITCH packet, it sends a MODIFY BEARER REQUEST to the SMF.
9. The SMF, on receiving the MODIFY BEARER REQUEST, provides an acknowledgment as a MODIFY BEARER RESPONSE packet to the AMF.
10. On receiving the MODIFY BEARER RESPONSE from the SMF, AMF acknowledges the Path switch request sent by the target gNB by sending the PATH SWITCH ACK packet back to the target gNB via Switch 5.
11. The target gNB sends a CONTEXT RELEASE packet to the source gNB, and the source gNB sends back a CONTEXT RELEASE ACK packet to target gNB. The context release request and ack packets are sent between the source and target gNB via Switch 6.
12. RRC Reconfiguration will take place between the target gNB and UE_9.
13. The UE_9 will now start sending the UE MEASUREMENT REPORT to gNB_8.

10.3.5 Plot of SNR Versus Time

The SNR variation as a function of time during the handover process is shown in Fig. 10.4; this plot can be obtained from the LTENRLog file. However, it involves first obtaining the values using Pivot tables, as explained in the previous experiments.

- At time 15.60 s, the SNR from gNB7 is 7.81 dB and the SNR from gNB8 is also 7.81 dB. This represents the point where the two curves intersect.
- Time 18.6 s, the SNR from gNB7 is 6.18 dB and the SNR from gNB8 is 9.51 dB. This represents the handover point where Adj cell RSRP is greater than serving cell RSRP by the hand-over-margin (HOM) of 3 dB.

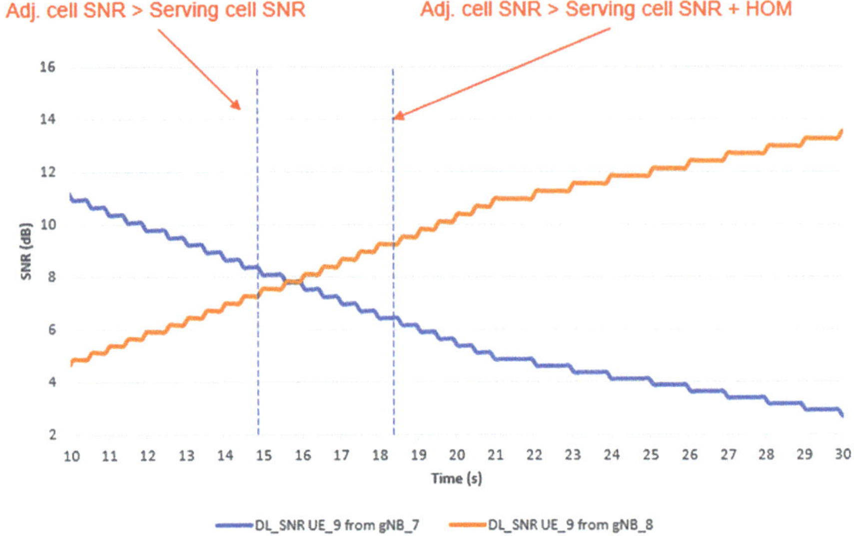

Fig. 10.4 Plot of DL SNR (at UE_3 from gNB1 and gNB2) versus time. The handover commences when Adj_cell_SNR > Serving_cell_SNR + Hand_over_Margin

10.4 Part II: Throughput and Delay Variation During Handover

10.4.1 Network Scenario

NetSim UI displays the configuration file corresponding to this experiment, as shown in Fig. 10.5.

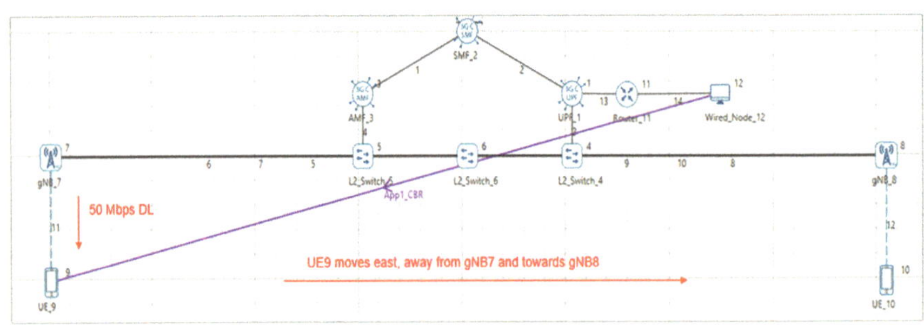

Fig. 10.5 Network set up for studying the throughput and delay variation during handover

Table 10.4 gNB_7 > Interface (5G_RAN) properties setting

Interface (5G_RAN) properties	
CA_Configuration	n78
CA_Count	1
Numerology	0
Channel bandwidth (MHz)	10
PRB count	52
MCS table	QAM64
CQI table	Table 1
X_Overhead	XOH0
DL UL ratio	4:1
Pathloss model	3GPPTR38.901-7.4.1
Outdoor scenario	Urban macro
LOS_NLOS selection	User_Defined
LOS probability	1
Shadow fading model	None
Fading_and_Beamforming	NO_FADING_MIMO_UNIT_GAIN

10.4.2 Network Configuration

1. The device positions are set as per Table 10.3.
2. The properties of gNB 7 are chosen as per Table 10.4.
3. The properties of gNB 8 are set similarly.
4. The transmit and receive antenna counts are set to 2 and 1, respectively, at the gNBs.
5. The transmit and receive antenna counts are set to 2 and 1, respectively, at the UEs.

10.5 Sample Results and Discussion

10.5.1 UDP Throughput Plot

As shown in the throughput plot in Fig. 10.6, the application starts at 1 s. The packet generation rate is 50 Mbps, and we see the network is able to handle this load, and the throughput is equal to the generation rate. We then observe that the throughput starts dropping from 2.5 s onwards because the UE is moving away from the gNB. As it moves further away from the serving gNB, the SNR falls, and therefore, a lower MCS is chosen, leading to reduced throughput (Molisch 2012). At 3 s, there is a further drop in throughput and then a final dip at 3.9 s. The time at which the handover occurs

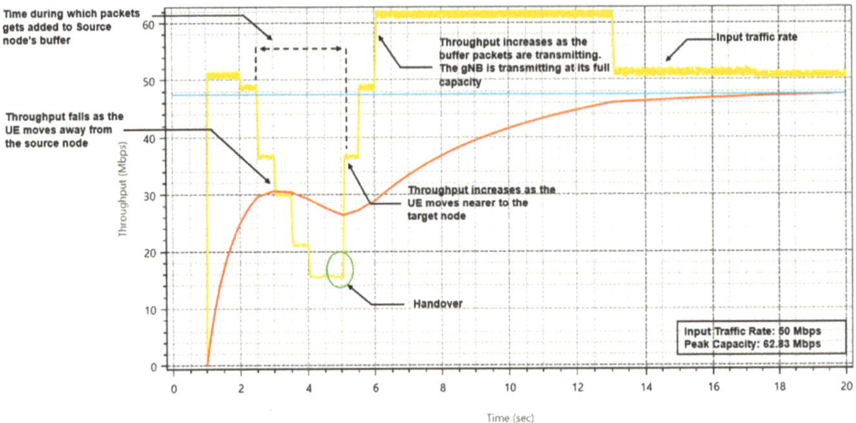

Fig. 10.6 Throughput variation versus time as the UE moves from the source gNB to the target gNB. Yellow curve represents the instantaneous throughput, and red curve represents the low-pass filtered version of the instantaneous throughput as a function of time

is 5.04 s. At this point, we see the throughput starts increasing once UE attaches to gNB8. For a short period of time, the throughput is greater than 50 Mbps because of the transmission of queued packets in the s-gNB buffer which get transferred to the t-gNB buffer over the Xn interface.

10.5.2 UDP Delay Plot

As shown in the delay plot in Fig. 10.7, since the application starts at 1 s, the initial UDP delay is $\approx$ 1 ms, and hence the curve is seen as close to 0 on the *Y* axis. We then see that the packet delay starts increasing as the UE moves away from the gNB. This is because the link capacity drops as the CQI falls. The peak delay experienced shoots up to $\approx$ 1.1 s at $\approx$ 5.5 s when the handover occurs. Once the handover is complete, the delay starts reducing and returns to $\approx$ 1 ms. The reason is that as the UE moves closer to the gNB, its CQI increases, and hence the 5G link can transmit at a higher rate, clearing all the buffered packets and restoring the delay to its initial value.

10.6 Exercises

1. Repeat the above experiment for a different transmit power. Explain your results.

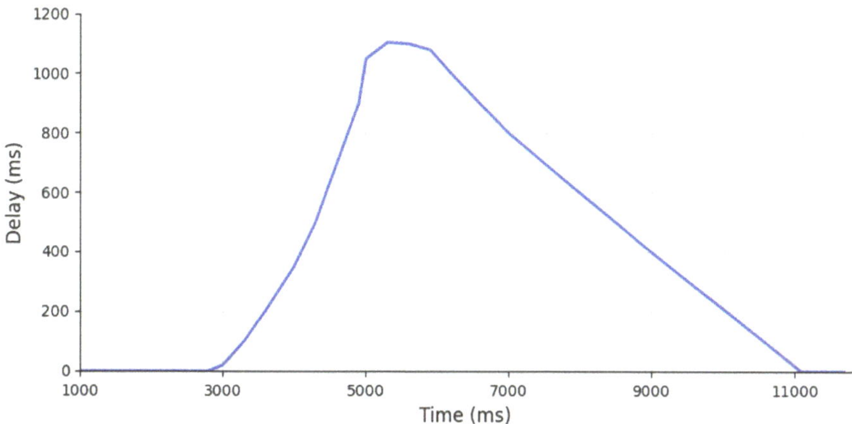

Fig. 10.7 Plot of delay versus time during handover

2. Understand the effect of handover margin on the process of handover by varying the handover margin from 1 to 10 dB. Interpret and justify all your results.

Appendix

Setting up the Experiment Configuration Files

Part I: Handover Algorithm

Open NetSim and click on **Experiments > 5G NR > Handover in 5GNR > Handover Algorithm**, then click on the tile in the middle panel to load the example shown in Fig. 10.8.

Part II: Throughput and Delay Variation During Handover

First, open the relevant configuration file by selecting **Experiments > 5G NR > Handover in 5GNR > Throughput and delay variation during handover > Effect of handover on Delay and Throughput**. See Fig. 10.9.

Setting the Network Configuration and Executing the Experiment for Part I: Handover Algorithm

The following procedures must be done to generate the sample results shown in this experiment.

Step 1: A network scenario is designed in NetSim GUI comprising a 5G Core, 2 gNBs, and 2 UEs in the **"5G NR"** Network Library.

Step 2: The device positions are set as per Table 10.1.

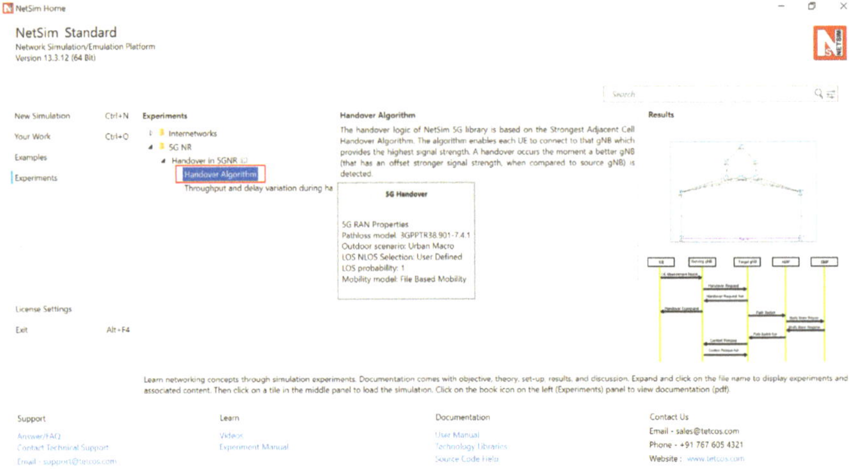

Fig. 10.8 List of scenarios for the example of handover in 5G NR

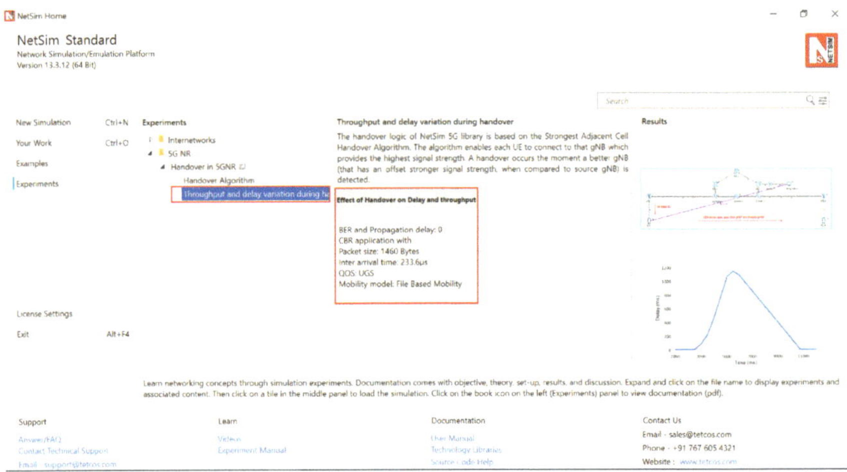

Fig. 10.9 List of scenarios for the example of handover in 5G NR for PART-2

Step 3: In the General Properties of UE_9 set Mobility Model as File-Based Mobility (see below).

Step 4: Right-click on gNB_7 and select Properties; set the properties as per Table 10.2. Similarly, it is set for gNB_8.

Step 5: The Tx_Antenna_Count is set to 2, and Rx_Antenna_Count is set to 1 in gNB > Interface (5G_RAN) > Physical Layer.

Step 6: The Tx_Antenna_Count is set to 1, and Rx_Antenna_Count is set to 2 in UE > Interface (5G_RAN) > Physical Layer.

Step 7: Right-click on the Application Flow **App1 CBR** and select Properties or click on the Application icon present in the top ribbon/toolbar.

A CBR Application is generated from UE 9 (Source) to UE 10 (Destination) with Packet Size 1460 Bytes and Inter Arrival Time 20,000 μs. QoS is set to UGS.

Additionally, the **"Start Time(s)"** parameter is set to 40 while configuring the application.

File-Based Mobility

In file-based mobility, users can write their own custom mobility models and define the movement of the mobile users. Create a mobility.txt file for UEs involved in mobility, with each step equal to 0.5 s, covering a distance of 50 m in each step.

Once imported, the NetSim Mobility File (mobility.txt) appears as an Excel sheet, which looks similar to that shown in Fig. 10.10.

Step 8: Packet Trace is enabled in NetSim GUI. At the end of the simulation, a very large .csv file containing all the packet information is available for the users to perform packet-level analysis. Ensure that plots are enabled in the NetSim GUI.

	A	B	C	D	E	F
1	#Time(s)	Device ID	X	Y	Z	
2	0	9	500	3000	0	
3	0.5	9	1000	3000	0	
4	1	9	1050	3000	0	
5	1.5	9	1100	3000	0	
6	2	9	1150	3000	0	
7	2.5	9	1200	3000	0	
8	3	9	1250	3000	0	
9	3.5	9	1300	3000	0	
10	4	9	1350	3000	0	
11	4.5	9	1400	3000	0	
12	5	9	1450	3000	0	
13	5.5	9	1500	3000	0	
14	6	9	1550	3000	0	
15	6.5	9	1600	3000	0	
16	7	9	1650	3000	0	
17	7.5	9	1700	3000	0	
18	8	9	1750	3000	0	
19	8.5	9	1800	3000	0	
20	9	9	1850	3000	0	
21	9.5	9	1900	3000	0	
22	10	9	1950	3000	0	
23	10.5	9	2000	3000	0	
24	11	9	2050	3000	0	
25	11.5	9	2100	3000	0	
26	12	9	2150	3000	0	
27	12.5	9	2200	3000	0	
28	13	9	2250	3000	0	
29	13.5	9	2300	3000	0	
30	14	9	2350	3000	0	
31	14.5	9	2400	3000	0	
32	15	9	2450	3000	0	
33	15.5	9	2500	3000	0	

	A	B	C	D	E	F
33	15.5	9	2500	3000	0	
34	16	9	2550	3000	0	
35	16.5	9	2600	3000	0	
36	17	9	2650	3000	0	
37	17.5	9	2700	3000	0	
38	18	9	2750	3000	0	
39	18.5	9	2800	3000	0	
40	19	9	2850	3000	0	
41	19.5	9	2900	3000	0	
42	20	9	2950	3000	0	
43	20.5	9	3000	3000	0	
44	21	9	3050	3000	0	
45	22	9	3100	3000	0	
46	23	9	3150	3000	0	
47	24	9	3200	3000	0	
48	25	9	3250	3000	0	
49	26	9	3300	3000	0	
50	27	9	3350	3000	0	
51	28	9	3400	3000	0	
52	29	9	3450	3000	0	
53	30	9	3500	3000	0	
54	31	9	3550	3000	0	
55	32	9	3600	3000	0	
56	33	9	3650	3000	0	
57	34	9	3700	3000	0	
58	35	9	3750	3000	0	
59	36	9	3800	3000	0	
60	37	9	3850	3000	0	
61	38	9	3900	3000	0	
62	39	9	3950	3000	0	
63	40	9	4000	3000	0	
64						
65						

Fig. 10.10 Mobility file sample

Step 9: The log file can be enabled per the information provided in **Section 3.22** of the 5G-NR technology library document available in NetSim's documentation.

Step 10: Run the simulation for 50 s.

Setting the Network Configuration and Executing the Experiment for Part II: Throughput and Delay Variation During Handover

The following set of procedures are done to generate this sample:

Step 1: A network scenario is designed in NetSim GUI comprising 2 gNBs, 5G Core, 1 Router, 1 Wired Node, and 2 UEs in the **"5G NR"** Network Library.

Step 2: The device positions are set as per Table 10.3.

Step 3: Right click on the gNB_7, select Properties, and set them as in Table 10.4. Similarly, set the properties for gNB_8.

Step 4: The Tx_Antenna_Count is set to 2 and Rx_Antenna_Count is set to 1 in gNB > Interface (5G_RAN) > Physical Layer.

Step 5: The Tx_Antenna_Count is set to 1 and Rx_Antenna_Count is set to 2 in UE > Interface (5G_RAN) > Physical Layer.

Step 6: In the General Properties of UE_9 and UE_10, set Mobility Model as File-Based Mobility (see below).

Step 7: The BER and propagation delay are set to zero in all the wired links.

Step 8: Right-click on the Application Flow **App1 CBR** and select Properties, or click on the Application icon present in the top ribbon/toolbar.

A CBR Application is generated from Wired Node 12 (Source) to UE_9 (Destination), with a Packet Size of 1460 Bytes and Inter Arrival Time of 233.6 μs. QoS is set to UGS.

Additionally, the **"Start Time(s)"** parameter is set to 1 while configuring the application.

File-Based Mobility

In file-based mobility, users can write their own custom mobility models and define the movement of the mobile users. Create a mobility.txt file for UE's involved in mobility with each step equal to 0.5 s, covering a desired distance in each step.

The NetSim Mobility File (mobility.txt) looks like as shown in Fig. 10.11.

Step 9: Packet Trace and Event Trace are enabled in the NetSim GUI. At the end of the simulation, a very large .csv file containing all the packet information is available for the users to perform packet-level analysis. Plots are enabled in the NetSim GUI.

Step 10: The log file is populated as per the information provided in **Section 3.22** in the 5G NR technology library document available in the NetSim documentation.

Step 11: Run the simulation for 20 s.

	A	B	C	D	E	F
1	#Time(s)	Device ID	X	Y	Z	
2	0	9	500	1000	0	
3	0.5	9	750	1250	0	
4	1	9	1000	1500	0	
5	1.5	9	1250	1750	0	
6	2	9	1500	2000	0	
7	2.5	9	1750	2250	0	
8	3	9	2000	2500	0	
9	3.5	9	2250	2750	0	
10	4	9	2500	3000	0	
11	4.5	9	2750	2750	0	
12	5	9	3250	2250	0	
13	5.5	9	3500	2000	0	
14	6	9	3750	1750	0	
15	6.5	9	4000	1500	0	
16	7	9	4250	1250	0	
17	7.5	9	4500	500	0	
18						
19						

Fig. 10.11 Mobility file sample for Part II

Steps to Log the Results for Part I: Handover Algorithm Mechanism

1. After the simulation, open the Event trace file from the Metric table as shown in Fig. 10.12.

Fig. 10.12 ISD 500 m simulation result dashboard

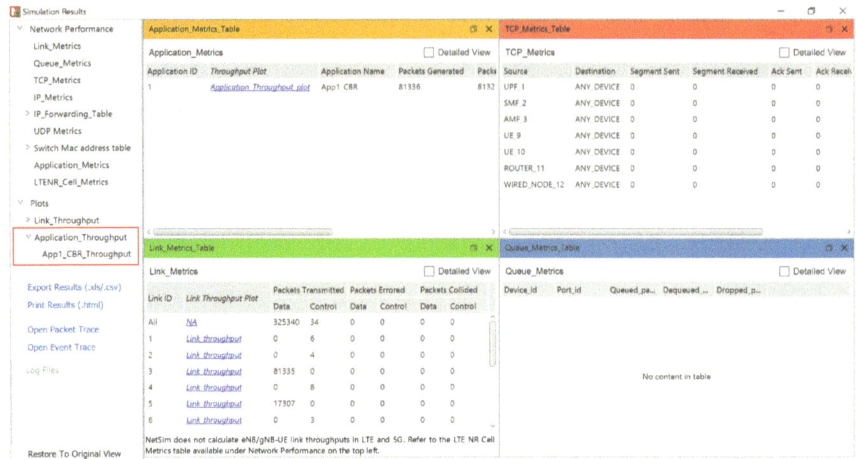

Fig. 10.13 Step to obtain the throughput plot

2. This yields the results shown in Fig. 10.3.
3. Next, to obtain the plot in Fig. 10.4 open the Log file, and follow the procedure given in the Appendix in Chap. 4.

Steps to Log the Results for Part II: Throughput and Delay Variation During Handover

Throughput Variation Plot

Obtain the plot by clicking on the "Application_Throughput" option under the "Plots" option, as shown in Fig. 10.13.

Delay Computation from Event Traces

NOTE: Follow the article link given below to generate a pivot table for large Packet Trace and Event Trace files

https://support.tetcos.com/en/support/solutions/articles/14000122911-how-to-generate-pivot-reports-for-large-packet-trace-and-event-trace-files

1. Open Event Trace after simulation.
2. Go to the **Insert** option at the top ribbon of the trace window and select **Pivot Tables**.
3. In the window that arises, you can see **Table_1**. Click on OK.
4. This will create a new sheet with a Pivot Table, as shown in Fig. 10.14.
5. Now, drag and drop **Packet_Id** to the **Rows** field. Similarly, drag and drop the following: **Event_Type** to **Columns** field, **Event_Time** to **Values** field, as shown in Fig. 10.15.

Fig. 10.14 Blank pivot table

Fig. 10.15 Adding fields into columns, rows, and values

6. Now, in the Pivot table formed, filter **Event_Type** to **APPLICATION_IN** and **APPLICATION_OUT** as shown in Fig. 10.16.
7. In the **Values** field in the Pivot Table Fields, Click on **Sum of Event Time (US)** and select **Value Field Settings** as shown in Fig. 10.17.
8. Select the **Show Values As** option and filter it to **Difference From**, as shown in Fig. 10.18.
9. In the **Base field**, select **Event_Type** and in the **Base_item** field select **APPLICATION_OUT** and click on OK, as shown in Fig. 10.19. This will provide the end-to-end delay in the pivot table.
10. Now, ignore the negative readings in the Delay values (Fig. 10.20) obtained and use the positive values to plot the Delay versus Time (APPLICATION_IN) graph.
11. In the Event trace window, filter the **Event Type** to **APPLICATION_IN** and use the Event Time thus obtained as the *x*-axis of the plot, as shown in Fig. 10.21.

NOTE: <u>To change the handover margin, do the following steps in NetSim.</u>

gNB Properties > Interface_4 (5G_RAN) > DATALINK_LAYER > HANDOVER > Handover Margin (dB)

	A	B	C	D	E
1					
2					
3	Sum of Event_Time(US)	Column Labels			
4	Row Labels	APPLICATION_IN	APPLICATION_OUT	Grand Total	
5	1	1001999	1000000	2001999	
6	2	1001999	1000292	2002291	
7	3	1001999	1000584	2002583	
8	4	1001999	1000876	2002875	
9	5	1002999	1001168	2004167	
10	6	1002999	1001460	2004459	
11	7	1002999	1001752	2004751	
12	8	1003999	1002044	2006043	
13	9	1003999	1002336	2006335	
14	10	1003999	1002628	2006627	
15	11	1003999	1002920	2006919	
16	12	1004999	1003212	2008211	
17	13	1004999	1003504	2008503	
18	14	1004999	1003796	2008795	
19	15	1006999	1004088	2011087	
20	16	1006999	1004380	2011379	
21	17	1006999	1004672	2011671	
22	18	1006999	1004964	2011963	
23	19	1006999	1005256	2012255	
24	20	1006999	1005548	2012547	
25	21	1007999	1005840	2013839	
26	22	1007999	1006132	2014131	

Sheet2 | Event Trace | Pivot Table(Custom)

Fig. 10.16 Event Type filtered to APPLICATION_IN and APPLICATION_OUT to calculate delay

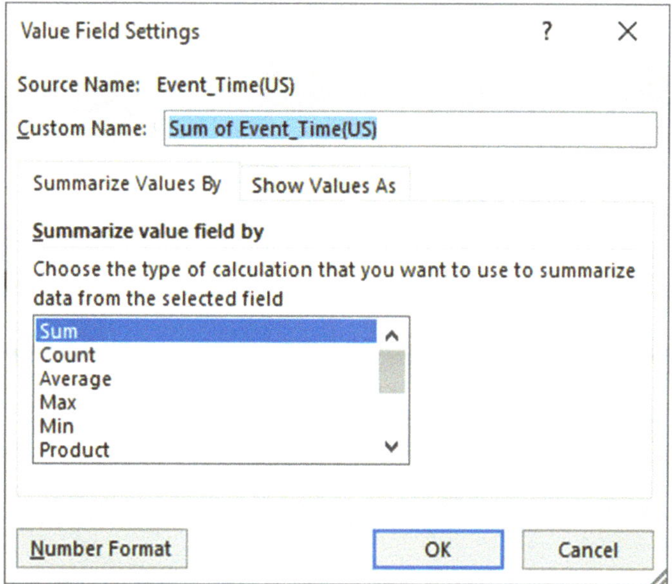

Fig. 10.17 Value field settings to Sum of Event Time

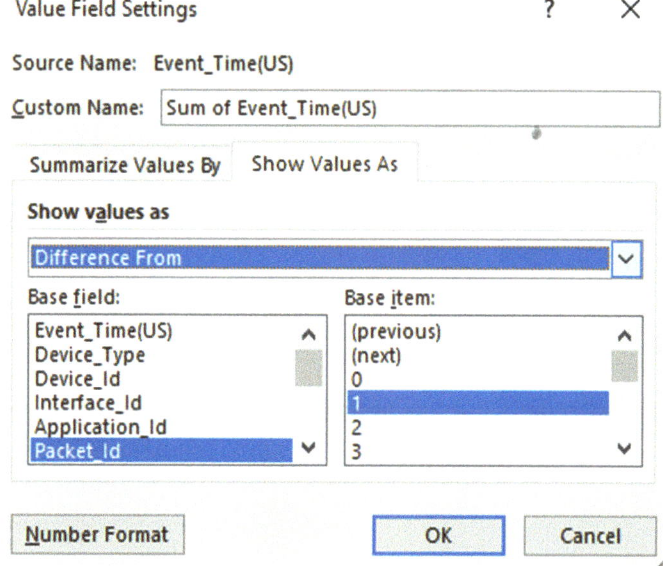

Fig. 10.18 Select Show Values as Difference From

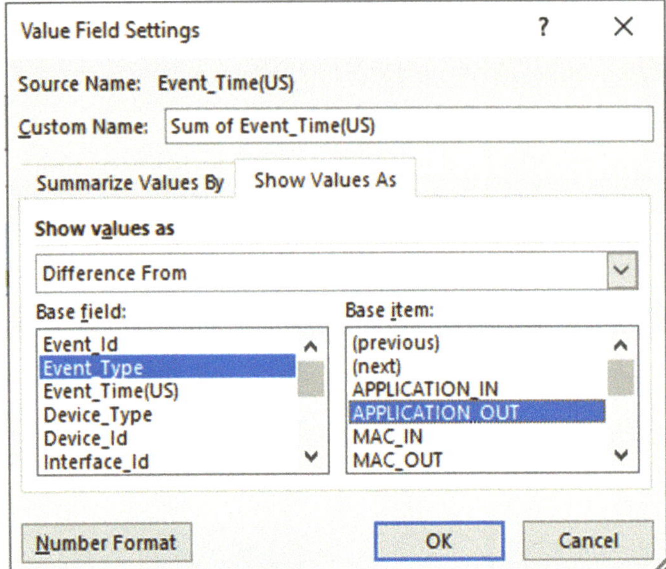

Fig. 10.19 Select Base field to Event Type and Base item to APPLICATION OUT

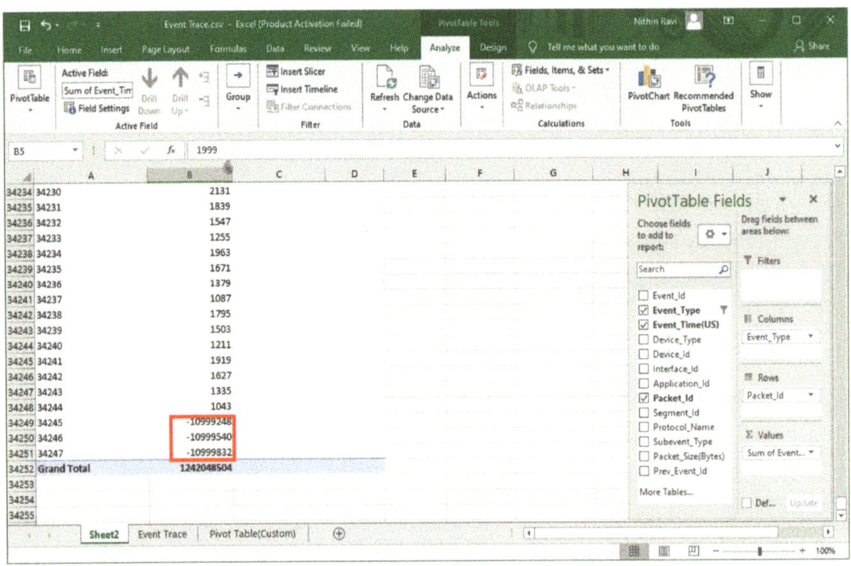

Fig. 10.20 Ignore the negative values in the delay

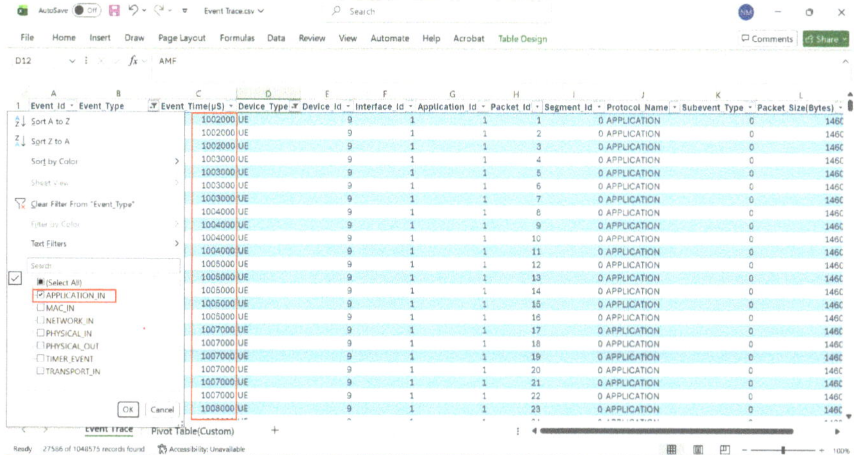

Event Time(µS)	Device Type	Device Id	Interface Id	Application Id	Packet Id	Segment Id	Protocol Name	Subevent Type	Packet Size(Bytes)
1002000	UE	9	1	1	1	0	APPLICATION	0	146C
1002000	UE	9	1	1	2	0	APPLICATION	0	146C
1002000	UE	9	1	1	3	0	APPLICATION	0	146C
1003000	UE	9	1	1	4	0	APPLICATION	0	146C
1003000	UE	9	1	1	5	0	APPLICATION	0	146C
1003000	UE	9	1	1	6	0	APPLICATION	0	146C
1003000	UE	9	1	1	7	0	APPLICATION	0	146C
1004000	UE	9	1	1	8	0	APPLICATION	0	146C
1004000	UE	9	1	1	9	0	APPLICATION	0	146C
1004000	UE	9	1	1	10	0	APPLICATION	0	146C
1004000	UE	9	1	1	11	0	APPLICATION	0	146C
1005000	UE	9	1	1	12	0	APPLICATION	0	146C
1005000	UE	9	1	1	13	0	APPLICATION	0	146C
1005000	UE	9	1	1	14	0	APPLICATION	0	146C
1005000	UE	9	1	1	15	0	APPLICATION	0	146C
1005000	UE	9	1	1	16	0	APPLICATION	0	146C
1007000	UE	9	1	1	17	0	APPLICATION	0	146C
1007000	UE	9	1	1	18	0	APPLICATION	0	146C
1007000	UE	9	1	1	19	0	APPLICATION	0	146C
1007000	UE	9	1	1	20	0	APPLICATION	0	146C
1007000	UE	9	1	1	21	0	APPLICATION	0	146C
1007000	UE	9	1	1	22	0	APPLICATION	0	146C
1008000	UE	9	1	1	23	0	APPLICATION	0	146C

Fig. 10.21 Event trace

References

Dimou K et al (2009) Handover within 3GPP LTE: design principles and performance. In: 2009 IEEE 70th vehicular technology conference fall, Anchorage, AK, USA, pp 1–5

Dutta A, Schulzrinne H (2014) Mobility protocols and handover optimization: design, evaluation and application. Wiley

Molisch AF (2012) Wireless communications, vol 34. Wiley

Chapter 11
On the Impact of MAC Schedulers on 5G NR Performance: Design and Analysis

11.1 Objective

In this experiment, our focus is to understand how different scheduling algorithms affect the UDP download throughput of a multi-user (multi-UE) system. We will focus on three popular scheduler algorithms: Max-rate scheduler, proportional fair (PF), and round-robin scheduler, and gain insights on the following aspects:

1. How does the throughput vary in multi-UE networks with different types of schedulers when the channel is not time-varying?
2. How does the throughput vary in multi-UE networks with different types of schedulers when the channel is time-varying?
3. How does the proportional fair scheduler work, and what is multi-user diversity?

11.2 Introduction and Theory

One of the key requirements of 5G systems is to support communication of many users with a wide range of devices and applications. These conditions give rise to heterogeneous traffic in the network. To carry such traffic in a wireless network, the design and development of schedulers capable of considering the channel conditions of each user are needed. Schedulers are usually the "secret sauce" to obtain superior performance of any network operator and hence are an important design element. Among the class of several schedulers, we will focus on three popular algorithms in this experiment: Max-rate, proportional fair, and round-robin (RR).

L. Yashvanth et al., *Understanding 5G New Radio*, Transactions on Computer Systems and Networks, https://doi.org/10.1007/978-981-92-0112-9_11

11.2.1 Max-Rate Scheduler

In this type of scheduler, the gNB schedules the UE, which observes the best instantaneous channel condition among all the UEs. Mathematically, in every time slot t, the user k^* is selected as per the following criterion:

$$k^*(t) = \arg \max_{k \in \{1,2,\ldots,K\}} |h_k(t)|^2,$$

where h_k is the channel seen by k th user from the gNB in a K-user system. Since the user with the best channel also obtains the best data rate among all users, the max-rate scheduler achieves the best possible total system throughput among all schedulers.

To understand the effect of max-rate scheduling on the impact of system throughput, we first understand the behavior of the max-rate scheduler's metric: $\max_{k \in \{1,2,\ldots,K\}} |h_k(t)|^2$ as a function of K, the total number of UEs in the system. As an example, if we consider that every UE experiences a Rayleigh channel from the gNB and assume that channels across UEs are independent, we can show that the probability distribution of the metric is as in Fig. 11.1.

From Fig. 11.1, as the total number of UEs increases, the best channel "hardens", i.e., it concentrates around its mean value, and the support of the distribution keeps shifting toward the right, indicating that the average SNR seen by the system under max-rate scheduling monotonically increases with the number of UEs in the system. The following theorem accurately quantifies this behavior.

Theorem 11.1 *Assume that a system is equipped with a single antenna gNB and K UEs with independent and identically distributed Rayleigh channels across the UEs. Then, under max-rate scheduling, the system spectral efficiency $R^{(K)}$ asymptotically (i.e., for large K) scales as*

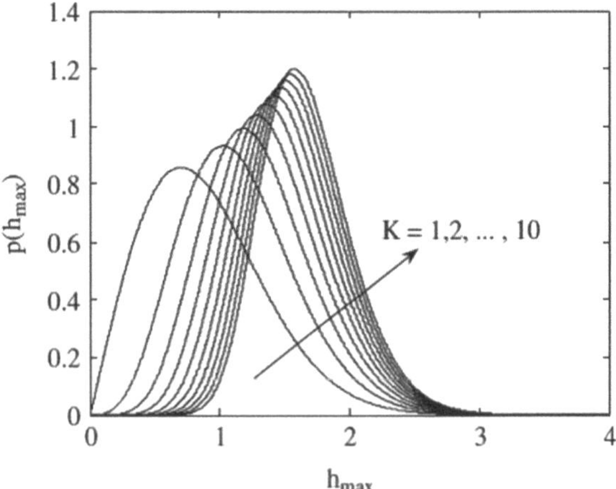

Fig. 11.1 Probability density function of $\max_{k \in \{1,2,\ldots,K\}} |h_k(t)|^2$.

$$\lim_{K\to\infty}\left(R^{(K)} - \log_2\left(1 + \frac{P}{\sigma^2} \times \log_e(K)\right)\right) = 0,$$

where P and σ^2 denote the transmit power and receiver noise power, respectively. ■

Thus, in a K-user system, the average SNR under max-rate scheduler scales as $\log_e(K)$, when max-rate scheduling is employed. This effect of obtaining additional gain in the SNR by having multiple (or many) users in the system and **opportunistically** scheduling the UEs in every time-resource element is called **multi-user diversity** (Tse and Viswanath 2005).

A naïve way to implement this type of scheduling strategy is to first broadcast a common pilot symbol (possibly in a mini slot) from the gNB to all the UEs, and each UE computes the instantaneous channel quality information (CQI). Then, all UEs feed back their CQI and the gNB schedules the UE with the best CQI. This process repeats in every time slot. There are more sophisticated ways of identifying the user with the best CQI that do not involve all UEs sending their feedback, however, these are not discussed here.

The max-rate scheduler achieves the highest possible system throughput and is therefore attractive to an operator who wants to maximize its revenue when the revenue depends mainly on the amount of data consumed by the UEs. However, a drawback with this approach is that this scheduler does not promote fairness in the scheduling of UEs. For instance, a UE that is far away from the gNB always witnesses a weak channel and would, therefore, rarely be scheduled. Such a user may remain unhappy and switch away from the operator, resulting in a loss of revenue in the long term. The round-robin scheduler, discussed next, is at the opposite end of the spectrum: it is absolutely fair across the UEs but loses out on the network throughput.

11.2.2 Round-Robin (RR)

In this type of scheduling mechanism, the emphasis of the scheduler is to ensure fairness among UEs in scheduling. As discussed above, although the max-rate scheduler achieves the highest system throughput, it does not ensure fairness among users in terms of the relative number of times any UE is scheduled. For example, if a UE is located at the cell-edge, then it is highly unlikely that the UE will get scheduled at any time slot due to high path loss, which deteriorates the CQI of the UE. To tackle this problem, the RR scheduler completely disregards the instantaneous CQI and schedules the users one after the other, in a round-robin fashion. This ensures that every user gets equal resource allocation regardless of the instantaneous CQI. A toy illustration of the mechanism of RR scheduling is shown in Fig. 11.2, for a system with 4 users.

Clearly, there is no feedback required from the UE to gNB, unlike the max–rate scheduler.

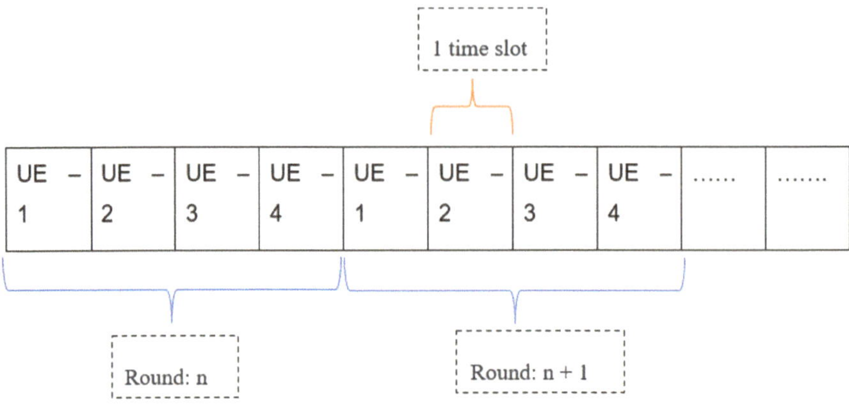

Fig. 11.2 Scheduling chart of round-robin scheduling

11.2.3 Proportional Fair (PF)

While the max-rate scheduler achieves the highest sum rate, it does not ensure fairness among users. On the other hand, the RR scheduler ensures the highest fairness among users, while the throughput of the system is compromised. This necessitates the need for a scheduler that strikes a balanced trade-off between the achievable throughput and fairness in the system. This is accomplished by the proportional-fair (PF) scheduler. In a PF scheduler, in any time slot t, a user k^* is selected as:

$$k^*(t) = \arg \max_{k \in \{1,2,\ldots,K\}} \frac{R_k(t)}{T_k(t)},$$

where $R_k(t) = \log_2\left(1 + \frac{|h_k(t)|^2 P}{\sigma^2}\right)$ and $T_k(t)$ are the instantaneous rate and average rate of user k till time t, respectively. In particular, $T_k(t)$ can be updated recursively as:

$$\mathrm{T_k(t+1)} = \begin{cases} \left(1 - \frac{1}{\mathrm{T_c}}\right) \mathrm{T_k(t)} + \frac{1}{\mathrm{T_c}} \mathrm{R_k(t)}, & \textit{if } k = \mathrm{k}^*(\mathrm{t}), \\ \left(1 - \frac{1}{\mathrm{T_c}}\right) \mathrm{T_k(t)}, & \textit{if } k \neq \mathrm{k}^*(\mathrm{t}). \end{cases}$$

In the above, the parameter $T_c \in (1, \infty)$ dictates the controllable trade-off between prioritizing throughput over fairness in the system, and is a designer's choice. It can be shown that,

- Higher values of T_c make the PF scheduler perform like a max-rate scheduler,
- Lower values of $T_c (T_c \approx 1)$ make the PF scheduler perform like an RR scheduler.

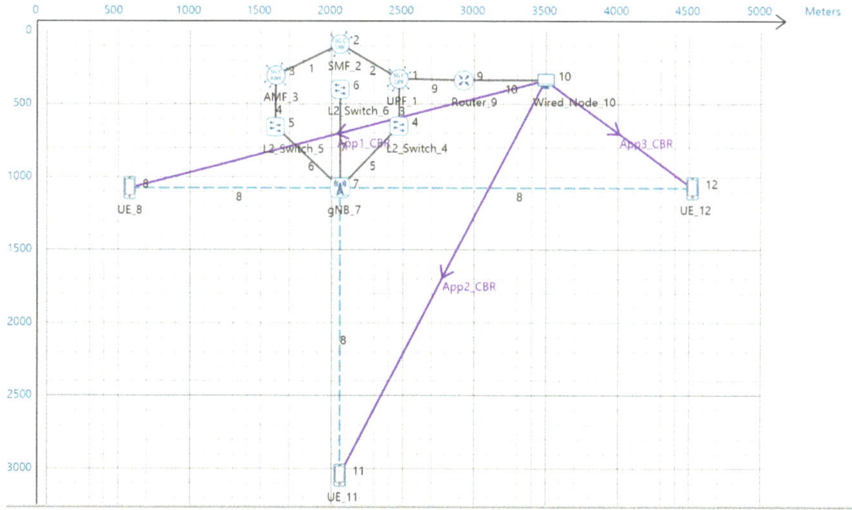

Fig. 11.3 Network set up for studying the scheduling example

11.3 Case I: UEs at Different Distances, and the Channel is Constant Over Time

11.3.1 Network Scenario

The network diagram in Fig. 11.3 illustrates the display of the NetSim UI when the example configuration file corresponding to this case is opened. See Appendix.

11.3.2 Network Configuration

1. Set the inter gNB-UE distances as follows:

$$\text{gNB\_7 to UE\_8} = 1500\text{ m},$$
$$\text{gNB\_7 to UE\_11} = 2000\text{ m},$$
$$\text{gNB\_7 to UE\_12} = 2500\text{ m}.$$

2. The gNB properties are set as per Table 11.1. In the first scenario, the scheduling type is set to Round Robin, in the second to proportional fair, and in the third to Max throughput.
3. The UE properties are set to the default values with single transmit and receive antennas.

Table 11.1 gNB properties

Properties	
Data link layer properties	
Scheduling type	Varies: proportional fair, Max throughput, Round Robin
Physical layer properties	
CA type	SINGLE_BAND
CA configuration	n78
CA1	
Numerology	1
Channel bandwidth	100 MHz
Outdoor_Scenario	URBAN_MACRO
LOS NLOS selection	USER_DEFINED
LOS probability	1
Pathloss model	3GPPTR38.901-7.4.1
Shadow fading model	None
Fading and beamforming	NO_FADING_MIMO_UNIT_GAIN

11.3.3 Results and Discussions

The results obtained when all three UEs (denoted by Application in the table below) simultaneously download the data are given in Table 11.2.

Next, consider a scenario in which only one of the UEs observes full DL traffic. First, run for the UE at 1500 m, and then for the UE at 2000 m, and then at 2500 m. This gives the maximum achievable throughput per node since the gNB resources (such as bandwidth) are not shared between 3 UEs and are fully dedicated to just one UE. The results for this case are given in Table 11.3.

Recall that the rate achieved at the PHY layer is decided by the received SNR. Therefore, a UE closer to the gNB will get a higher data rate than a farther UE. In this example, note that the distances from the gNB are such that UE_12 Distance > UE_11 Distance > UE_8 Distance. As a consequence, the throughputs are in the order UE_8 > UE_11 > UE_12.

Table 11.2 UDP download throughputs for different scheduling algorithms when all three UEs are simultaneously downloading data

Throughput (Mbps)				
Scheduling	Application 1	Application 2	Application 3	Aggregate
Round Robin	63.93	36.67	16.39	116.99
Proportional fair	63.93	36.67	16.39	116.99
Max throughput	191.36	0	0	191.36

Table 11.3 UE throughputs if they run standalone (without the other UEs downloading data)

Distance from gNB (m)	Application ID	Throughput (Mbps)	Remarks
1500	1	191.36	UE 8 alone has full buffer DL traffic
2000	2	110.16	UE 11 alone has full buffer DL traffic
2500	3	49.266	UE 12 alone has full buffer DL traffic

Now, in Round-Robin scheduling, since PRBs are allocated equally among the three UEs, the individual throughputs seen by each of the UEs are exactly $\frac{1}{3}$ the throughputs shown in Table 11.5. The PF scheduler results match those of the RR scheduler because the channel is not time-varying. Finally, in Max-throughput scheduling, since the PRBs are allocated such that the system gets the maximum download throughput, the nearest UE will always get all the resources, and so its throughput will be 3 times the throughput obtained under RR scheduling while the other UEs get a throughput of zero.

11.4 Case II: UEs at Different Distances and the Channel is Time-Varying

The procedure and configuration remain the same as the first case, except that now we will enable the fading in the settings to obtain temporal variations in the channel. Specifically, we consider Rayleigh channels in all the links of the system.

11.4.1 Results and Discussion

The achieved throughputs at the UEs for different scheduling algorithms when the channels vary with time are given in Table 11.4.

Table 11.4 UDP download throughputs for different scheduling algorithms when all three UEs simultaneously downloading data

Throughput (Mbps)				
Scheduling	Application 1	Application 2	Application 3	Aggregate
Round Robin	49.80	30.20	18.78	98.78
Proportional fair	64.87	38.40	23.12	126.39
Max throughput	131.61	31.69	0	163.30

Table 11.5 gNB > interface (5G_RAN) > data link/physical layer properties

Properties	
Data link layer properties	
Scheduling type	Proportional Fair
EWMA averaging rate ($\alpha = T_c$)	Vary: 1.001, 50, 9999
Physical layer properties	
Fading and beamforming	RAYLEIGH_WITH_EIGEN_BEAMFORMING

- When the channel is time-varying, the RR scheduler yields lower throughput than the PF scheduler. This is because the RR scheduler does not take advantage of the knowledge that a UE has a good channel and continues to serve the UEs cyclically, irrespective of the channel state of the UEs.
- The performance of the proportional-fair scheduler has improved over the case when the channel is not time-varying. This is because, when the channel fluctuates, the multi-user diversity gain kicks in because of the variations in the channel coefficients while performing opportunistic selection of users.
- As expected, the system throughput performance of the max-rate scheduler is better than the other two. However, we notice that the obtained rate under the max-rate scheduler when the channel varies is lower than in Case I, where the channels are static. This is because of the concavity of the rate as a function of the SNR—in layman's terms, poor channel states decrease the average throughput more than the increase brought about by good channel states. Hence, the sum-rate is lower than in Case I.

11.5 Case III: A Study of the Performance of the PF Scheduler

In this study, we will understand the design of the proportional fair scheduler in detail. Specifically, we will study the performance of the PF scheduler when the factor, T_c as it appears in the average throughput calculation, is varied. Note that, while using the PF schedulers in Cases I and II, we had set $T_c = 50$. In NetSim, T_c is called the "EWMA Averaging rate," where EWMA stands for the "exponentially weighted moving average".

1. As shown in Table 11.5, set $T_c = 1.001$, 50, and 9999 successively, and obtain the throughputs of the three UEs under PF scheduler.
2. The rest of the procedure and settings remain the same as in Case II.
3. Run the simulation for 1.5 s and note the throughput from the results window.

Table 11.6 UDP download throughputs for PFS scheduling when all three UEs are simultaneously downloading data

Throughput (Mbps)				
Scheduling using PFS	Application 1	Application 2	Application 3	Aggregate
$\alpha = 1.001$	54.40	30.95	16.86	102.21
$\alpha = 50$	64.87	38.40	23.12	126.39
$\alpha = 9999$	69.93	42.42	23.38	135.73

11.5.1 Results and Discussion ($\alpha = T_c$)

The achieved UE throughputs for different values of α under PF scheduling are given in Table 11.6.

We make the following observations:

- In all cases, the throughputs of UEs vary as UE_8 > UE_11 > UE_12. This is because of the increasing path loss of UEs in the order: UE_8 < UE_11 < UE_12
- At any given UE, as α is varied, the system throughput varies as: $\alpha = 9999 > \alpha = 50 > \alpha = 1.001$. That is, the system throughput increases with α. This can be explained as follows: We first note that α roughly captures the length of the time window over which we desire the scheduler to promote fairness in scheduling of UEs. Thus, for smaller values of α, the scheduler ensures fairness over a short term, and this leads to the throughput of the proportional-fair scheduler to behave like that of a round-robin scheduler. On the other hand, for large values of α, the scheduler prefers to "opportunistically" schedule UEs whose channel quality is better than the other UEs while maintaining a long-term fairness, which results in a performance close to that obtained under a max-rate scheduler. At moderate values of α, the scheduler chooses a UE that obtains the best channel gain relative to its average throughput witnessed till then for data transmission, but the shorter averaging window compared to the case with high value of α forces the scheduler to maintain some amount of short-term fairness too. Thus, the throughput in this case is better than a round-robin scheduler, while still smaller than a max-rate scheduler.
- Thus, the choice of α is crucial to the performance of a proportional-fair scheduler in terms of striking a balance between the system throughput and the time duration over which fairness is ensured in the scheduling of UEs.

11.5.2 Exercises

1. The first exercise is to understand the working of round-robin, max-rate, and PF schedulers, in an asymmetrical scenario, with the inter-gNB distances set as follows:

$$\text{gNB} - \text{UE\_1 distance} = 1400 + (2 * \text{x});$$
$$\text{gNB} - \text{UE\_2 distance} = 1900 + (2 * \text{x});$$
$$\text{gNB} - \text{UE\_3 distance} = 2400 + (2 * \text{x});$$

where x is an arbitrary two-digit number. Replicate the experiment with these distance values and justify the differences in the results.

2. In this exercise, we will study the effect of multi-user diversity by varying the number of transmit antennas at the gNB and observe the throughputs. For this experiment, consider only the performance of the max-rate and PF schedulers. Further, consider all UEs placed at the same distance from the gNB, with inter gNB- UE distance = 1400 + (2*x);

Vary the transmit antenna count as 1, 4, and 128 at the gNB only and study the throughput (where fading is enabled) for each antenna count. Infer your results and justify.

3. In this exercise, we will demonstrate the utility of multi-user diversity as a function of the number of UEs in the system. Consider a symmetrical scenario such that inter gNB-UE distance = 1400 + (2*x). Enable fading in the system (i.e., similar to Case II of this experiment.)

3a. First place 2 UEs in the system, and obtain the throughputs of max-throughput scheduler, PFS with $\alpha = 1.1$, PFS with $\alpha = 100$, PFS with $\alpha = 9999$. Compute the aggregate throughputs in each of the cases.

3b. Now repeat the above exercise when there are 4, 5, 8, and 10 UEs in the system with four aggregate throughputs (for max-throughput scheduler, PFS with $\alpha = 1.1$, PFS with $\alpha = 100$, PFS with $\alpha = 9999$) for each of 4 –, 5 –, 8 –, and 10 – UE systems.

3c. Report all your values in a tabular form as given in Table 11.7.

3d. Now plot a graph showing the aggregate throughput (y-axis) as a function of number of UEs (x-axis) for different scheduling schemes (different curves).

3e. Infer and justify all your results.

Hint:

Read Sec. II of the reference (Yashvanth and Murthy 2023), and you might want to produce plots like Fig. 1 of this paper.

Table 11.7 Aggregate throughput

Aggregate throughput (Mbps) in symmetrical scenario					
Scheduler	2–UE system	4–UE system	5–UE system	8–UE system	10–UE system
Max–rate	To be filled	To be filled	To be filled	To be filled	To be filled
PFS; $\alpha = 1.1$	To be filled	To be filled	To be filled	To be filled	To be filled
PFS; $\alpha = 100$	To be filled	To be filled	To be filled	To be filled	To be filled
PFS; $\alpha = 9999$	To be filled	To be filled	To be filled	To be filled	To be filled

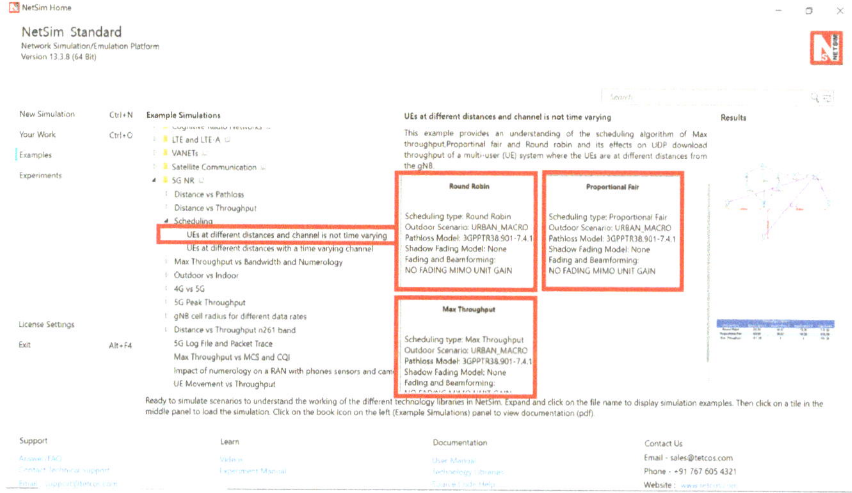

Fig. 11.4 List of scheduling scenarios for a non-time-varying channel

Note: In the above paper, τ denotes the PF scheduler constant, which equals α in this experiment.

Appendix

We provide the NetSim implementation details of this experiment in this appendix. The details are specific to NetSim v13.3.

Case I

Setting up the Configuration Files and Executing the Experiment.

Open NetSim, Select **Examples - > 5G NR - > Scheduling - > UEs at different distances and channel is not time-varying**. Then click on one of the tiles (as required) in the middle panel to load the example as shown in Fig. 11.4.

Configuring the scheduling algorithm and parameter settings in the example config files

1. Set the grid length to 5000 m in the Environment settings.
2. Set the distance values as follows.
   (a) gNB_7 to UE_8 = 1500 m
   (b) gNB_7 to UE_9 = 2000 m, and
   (c) gNB_7 to UE_10 = 2500 m
3. Go to the Wired link properties and set the properties as shown in Table 11.8.

Table 11.8 Wired link properties

Wired link properties	
Uplink speed	10,000 Mbps
Downlink speed	10,000 Mbps

4. Go to gNB properties - > 5G Interface (5G_RAN), and set the properties as per Table 11.1.
5. Go to Application Properties and set the properties as shown in Table 11.9.
6. Ensure that the plots are enabled in the NetSim GUI.
7. Run the simulation for 1.5 s and note down the throughput value from the results window in each sample. Recall that each scenario has a different scheduling algorithm configured.

Case II

Open NetSim, Select **Examples - > 5G NR - > Scheduling - > UEs at different distances with time-varying channel**. Then click on one of the tiles (as required) in the middle panel to load the example as shown in Fig. 11.5.

The rest of the procedure and configuration remain the same as above in the first case, except that now we will enable fading to obtain temporal variations in the channel, as shown in Table 11.10.

Run the simulation for 1.5 s and note down the throughput value from the results window.

Case III

Open NetSim, Select **Examples - > 5G NR - > Scheduling - > UEs at different distances with time-varying channel**. Then click on the "**Proportional Fair**" tile in the middle panel to load the example as shown in Fig. 11.6.

1. Open the data link layer properties in the 5G RAN of the gNB interface.

Table 11.9 Application properties

Application properties			
	Application 1	Application 2	Application 3
Application type	CBR	CBR	CBR
Source ID	10	10	10
Destination ID	8	9	10
QoS	UGS	UGS	UGS
Transport protocol	UDP	UDP	UDP
Packet size	1460 Bytes	1460 Bytes	1460 Bytes
Inter-arrival time	10 μs	10 μs	10 μs
Start time	1 s	1 s	1 s

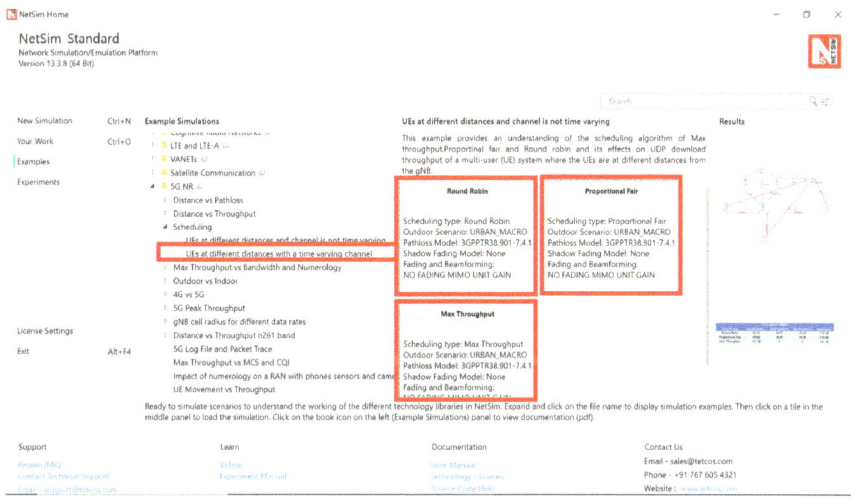

Fig. 11.5 List of scheduling scenarios for a time-varying channel

Table 11.10 Data link layer properties

Properties	
Data link layer properties	
Scheduling type	Varies: Proportional Fair, Max throughput, Round Robin
Physical layer properties	
Fading and beamforming	RAYLEIGH_WITH_EIGEN_BEAMFORMING

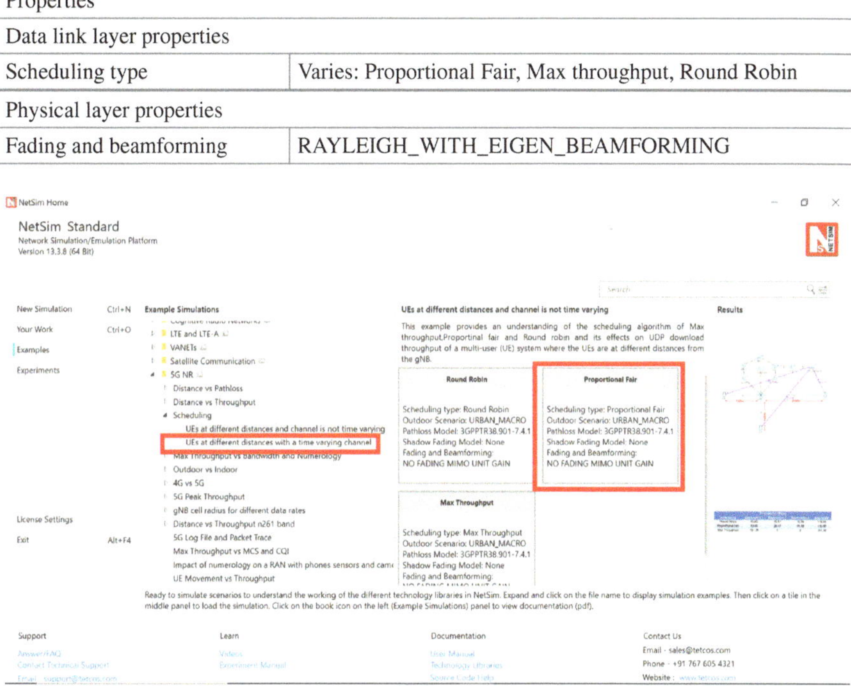

Fig. 11.6 Option to open the proportional-fair scheduler under a scenario with a time-varying channel

2. As shown in Table 11.5, set $\alpha = 1.001, 50$, and 9999 successively, and obtain the throughputs of the three UEs under the PF scheduler with each of these three values of α.
3. The rest of the procedure and settings remain the same as in Case II.
4. Run the simulation for 1.5 s and note down the throughput value from the results window.

Steps to Log the Simulation Results

The results will appear in the Application Metrics table. There is no need for any log files to be opened in this experiment.

Procedure to Insert New UEs in the Network (for Exercise 3)

1. Click on the UE icon in the toolbar present at the top of the NetSim UI and left-click the mouse again on the grid somewhere near the gNB to insert the UE. You can change the coordinates of the UE to achieve an inter-gNB UE distance as desired.
2. Establish a connection between the gNB and UE by clicking on the option "Wired/ wireless" on the top toolbar, and connect the gNB and the new UE by placing the mouse at the gNB, then dragging it, and clicking the mouse again after placing the mouse at the UE.
3. Then, to add a new application, click on the "Application" option on the top toolbar, and create a new application. To do this, in the application box, on the left, there will be an option to create a new application (look at the " + " symbol). After creating one, change the source node to the device ID of the "Wired Node" icon in the GUI; maybe this is device ID_13—this needs to be looked up. Then select the destination ID as the device ID of the newly inserted UE.
4. Ensure that all the "Application properties" and "UE properties" of the newly inserted UE are identical to the existing UEs for fair comparison.
5. Repeat this procedure for all the UEs that need to be newly inserted in the system.

References

Tse D, Viswanath P (2005) Multiuser capacity and opportunistic communications. In: Fundamentals of wireless communications, Cambridge University Press.

Yashvanth L, Murthy CR (2023) Performance analysis of intelligent reflecting surfaces assisted opportunistic communications. [Online]. Available: arxiv.org/abs/2203.06313.

BY

Chapter 12
An Overview of the Performance of Large MIMO Systems in 5G: A Theoretical Study

12.1 Objective

In this experiment, properties of MIMO channel matrices in 5G wireless communications are studied with a large number of antennas. In particular, the condition number of a large MIMO matrix, which dictates the performance of spatial multiplexing and, subsequently, the behavior of the eigen spectrum of large MIMO matrices, is investigated through a simulation setup in NetSim v13.3.

12.2 Introduction

A MIMO channel matrix in wireless communications is, in general, a random matrix. Hence, all the parameters obtained using the entries of this matrix are essentially random variables. Accordingly, the eigenvalues, the condition number are all random variables. In particular, the condition number of a MIMO channel matrix is important in characterizing the capacity of a MIMO channel. For example, by virtue of Jensen's inequality, we can bound the achievable throughput from the MIMO system as

$$\sum_{i=1}^{r} \log\left(1 + \frac{P_i \sigma_i}{N_0}\right) \leq r \log\left(1 + \frac{1}{r} \sum_{i=1}^{r} \frac{P_i \sigma_i}{N_0}\right),$$

where r is the rank of the MIMO channel matrix, σ_i is the ith singular value of the channel, P_i is the power alloted to the ith eigen mode, and N_0 is the noise variance. Further, the equality holds if and only if the condition number (the ratio of the largest to smallest nonzero singular values) of the MIMO matrix is 1. Thus, for a given rank and SNR of the channel, the condition number of the matrix determines how close the achievable rate can be to the upper bound. In this view, the distribution of the eigen spectrum and condition number of large MIMO matrices are important and

L. Yashvanth et al., *Understanding 5G New Radio*, Transactions on Computer Systems and Networks, https://doi.org/10.1007/978-981-92-0112-9_12

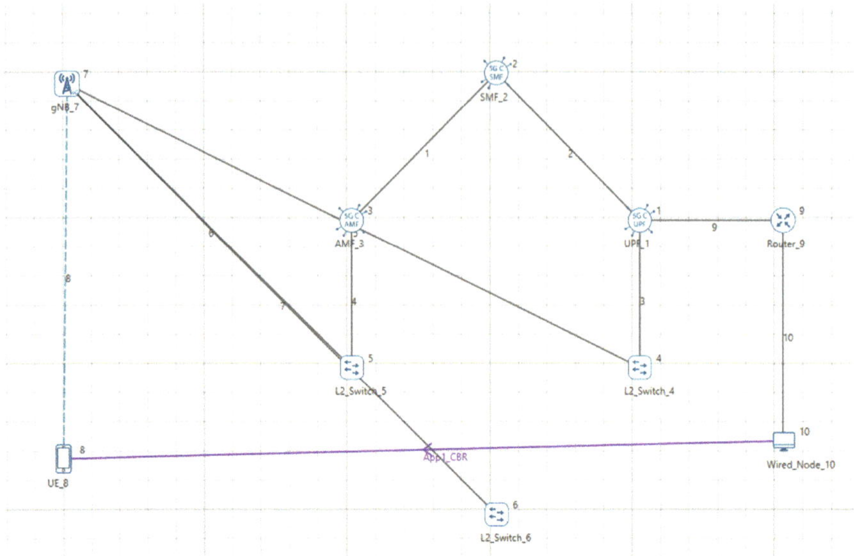

Fig. 12.1 Network topology in this experiment

are investigated in this experiment. By means of theory, it will be shown that the condition number of a large MIMO matrix stabilizes near unity, indicating superior spatial multiplexing capability at an asymptotic number of antennas.

12.3 Network Scenario

NetSim UI will display the network topology shown in Fig. 12.1 when you open the example configuration file.

12.4 Network Configuration

The following parameters are configured in the network setup:

1. The gNB-Interface 5G_RAN is set with the properties as shown in Table 12.1.
2. The UE properties are configured with the parameters as shown in Table 12.2.

Table 12.1 gNB properties

gNB-interface 5G_RAN parameters	
gNB height	10 m
Tx power	40 dBm
Duplex mode	TDD
CA type	SINGLE BAND
CA configuration	n78
DL: UL ratio	4:1
Numerology	0
Channel bandwidth (MHz)	10
Tx antenna count	Varied from 16 to 128
Rx antenna count	16
MCS table	QAM64
CQI table	TABLE1
Pathloss model	3GPPTR38.901-7.4.1
Outdoor scenario	Rural Macro
LOS NLOS selection	User Defined
LOS probability	1 (LOS)
Shadow fading model	3GPPTR38.901-7.4.1
Fading and beam forming	RAYLEIGH with EIGEN Beamforming
Coherence time (ms)	10
Additional loss model	None

Table 12.2 UE properties

UE interface 5G RAN	
Tx power	23 dBm
UE height	1.5 m
Tx antenna count	16
Rx antenna count	16

12.4.1 Part I: Asymptotic Mean Condition Number Evaluation

Using different choices of the number of transmit antennas, in Table 12.3, we provide the values of the condition numbers of large MIMO channel matrices as evaluated by NetSim.

On the other hand, we recall the theoretical values for the condition numbers below, which are obtained by using results from random matrix theory. The expected value (or mean) of the condition number of a $N_r \times N_t$ MIMO channel in the asymptotic number of antennas is given by (Edelman (1988))

Table 12.3 Mean and variance of the condition number with varying number of Tx antennas

Tx, Rx	Condition number mean	Condition number variance
16, 16	50.122	3168.247
32, 16	4.423	0.259643
64, 16	2.561	0.026433
128,16	1.878	0.005593

$$K = \frac{1 + \sqrt{\frac{n}{N}}}{1 - \sqrt{\frac{n}{N}}},$$

where $n = \min(N_r, N_t)$ and $N = \max(N_r, N_t)$. Then, we can compute the theoretical values of the condition number for different choices as considered in Table 12.3 in the following.

(a) Case1: gNB Tx = 32, UE Rx = 16:

From theory, $K = \frac{\left(1+\sqrt{\frac{16}{32}}\right)}{1-\sqrt{\frac{16}{32}}} = 5.82$, and the experimental result from NetSim = 4.423.

Therefore, difference $= \frac{5.82-4.423}{4.423} = 31.58\%$.

(b) Case2: gNB Tx = 64, UE Rx = 16:

From theory, $K = \frac{\left(1+\sqrt{\frac{16}{64}}\right)}{1-\sqrt{\frac{16}{64}}} = 3.0$, and the experimental result from NetSim = 2.561. Therefore, difference $= \frac{3-2.561}{2.561} = 17.1\%$.

(c) Case3: gNB Tx = 128, UE Rx = 16:

From theory, $\mathrm{K} = \frac{\left(1+\sqrt{\frac{16}{128}}\right)}{1-\sqrt{\frac{16}{128}}} = 2.09$, and the experimental result from NetSim = 1.878. Therefore, difference $= \frac{2.09-1.878}{1.878} = 11.28\%$.

We note that the difference between the experimental and theoretical values of the condition number of the channel matrix is not negligible. However, the theoretical values are applicable when there are a very large number of antennas, whereas the number of antennas in our experiments on NetSim is relatively modest. We observe that, as N increases, the difference reduces, and simulation outputs approach theoretical predictions.

12.4.2 *Part II: Distribution of the Asymptotic Condition Number*

Consider a i.i.d. $N_r \times N_t$ complex Gaussian random matrix **H**. Define the Wishart matrix $\boldsymbol{W} = \boldsymbol{H}\boldsymbol{H}^H$ (or $\boldsymbol{H}^H\boldsymbol{H}$—see Chap. 4) with parameters $n = \min(N_t, N_r)$ and $N = \max(N_t, N_r)$, and eigenvalues as $\lambda_{\max} = \lambda_1 \geq \lambda_2 \geq \ldots \geq \lambda_{\min} \geq 0$.

Now, the condition number of **H** is defined as

$$K(H) = \sqrt{\frac{\lambda_{\max}}{\lambda_{\min}}}.$$

From (Edelman 1988), if $n = N$ and $N \to \infty$, we can show that $\frac{K(\boldsymbol{H})}{n}$ converges in distribution to a random variable whose probability density function (PDF) is given by

$$f(x) = \left(\frac{8}{x^3}\right) \times e^{-\left(\frac{4}{x^2}\right)}.$$

In Fig. 12.2, we simulate the distribution of the condition number for the scenario: $N_r = N_t = 16$, and we consider these dimensions because the antenna count at the UEs is limited to 16 in NetSim. The figure shows a comparison of the normalized histogram of $\frac{K(\boldsymbol{H})}{n}$ from NetSim vs. the asymptotic PDF equation given above. Clearly, even with $N_r = N_t = 16$, the theoretical and empirical distribution demonstrate a reasonably good match.

Histogram for 16 Tx Layer Count (gNB) and 16 Rx Layer Count (UE)

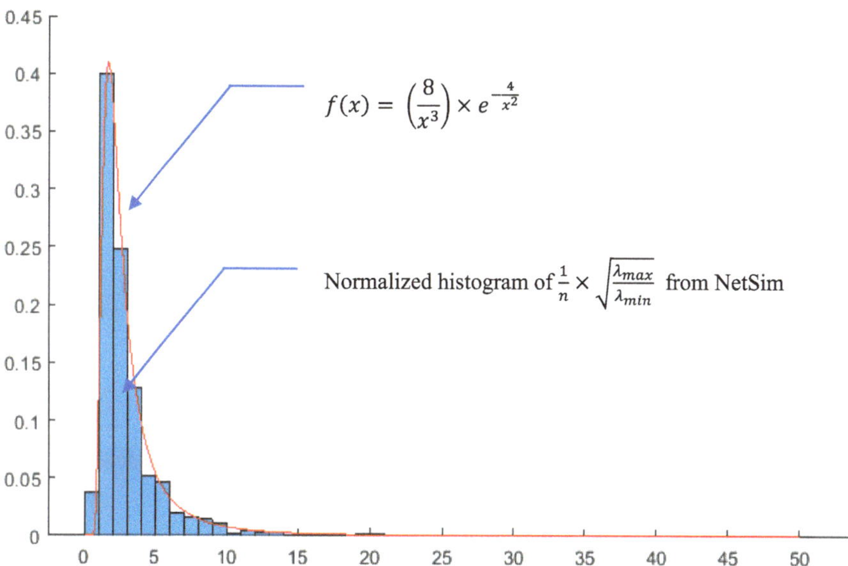

Fig. 12.2 The normalized histogram of with $N_t = N_r = 16$ fits well with the asymptotic distribution of $\frac{K(\boldsymbol{H})}{n}$

12.4.3 Part III: Marchenko-Pastur Distribution

This well-known distribution captures the joint distribution of the eigenvalues of a channel matrix. The Marchenko-Pastur (MP) distribution with $\frac{N_r}{N_t} = y$ and $N_t \to \infty$ is given by (Edelman (1988))

$$f(x) = \frac{1}{2\pi xy} \times \sqrt{(b-x)(x-a)},$$

where $b = \left(1+\sqrt{y}\right)^2$ and $a = \left(1-\sqrt{y}\right)^2$.

For example, consider the case where the number of transmit antennas N_t and the number of receive antennas N_r, are related as $\frac{N_r}{N_t} = \frac{1}{8} = y$. Substituting for y in the above equation, the MP distribution simplifies as

$$f(x) = \begin{cases} \frac{4}{\pi x}\sqrt{\left(\frac{1}{2}\right) - \left(x - \frac{9}{8}\right)^2}, & \left(1 - \frac{1}{2\sqrt{2}}\right)^2 \leq x \leq \left(1 + \frac{1}{2\sqrt{2}}\right)^2 \\ 0, \text{otherwise}. \end{cases}$$

Below in Fig. 12.3 we provide a sample result which shows that the empirical eigen spectrum of the Wishart matrix corresponding to the MIMO channel and the MP distribution above tightly match with each other. This validates the accuracy of the eigen spectrum characterization using the MP distribution in the large MIMO regime, which can then be used for subsequent analysis on the performance (e.g., multiplexing gain) of large MIMO systems.

12.5 Exercises

- Repeat part-3 of the experiment for $N_t = 16, 32$ also and analyze the fitness of the MP distribution for these cases.

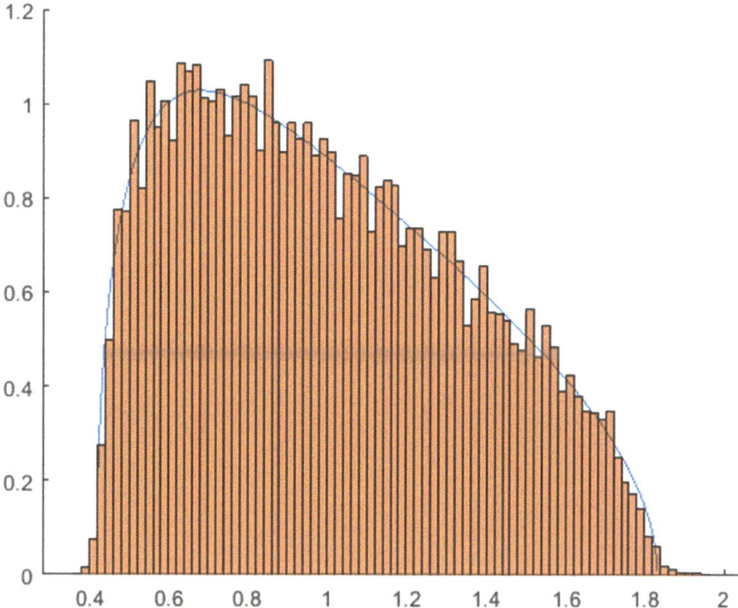

Fig. 12.3 NetSim results versus Marchenko-Pastur distribution for $N_r = 16$ and $N_t = 128$

Appendix

Procedure to Import the Workspace for the Experiment

1. Follow steps 1–8 of the Appendix in Chap. 4 and get to the stage as shown in Fig. 12.4.

Setting the Network Configurations and Executing the Experiment

1. The gNB and UE properties are set as given in Tables 12.1 and 12.2, using the physical layer option in the 5G RAN properties of the gNB/UE interfaces.
2. A downlink CBR application is configured from the wired node to the UE with transport protocol as UDP, packet size of 1460 bytes, and inter arrival time of 2000 μs, and the start time was set to 1 s.[1]
3. Click on the "Logs" icon in the toolbar to enable LTENR radio measurement as shown in Fig. 12.5.
4. Run simulation for 10 s. Note down the average linear beamforming gain (eigenvalue) obtained for the DL application from the log file generated, and then the condition number can be obtained as explained below.

[1] The application end time value of 10,000 s is not changed. In NetSim the application runs for min(AppEndTime, SimulationTime). Since the simulation is run for 10 s, the application runs for only 10 s.

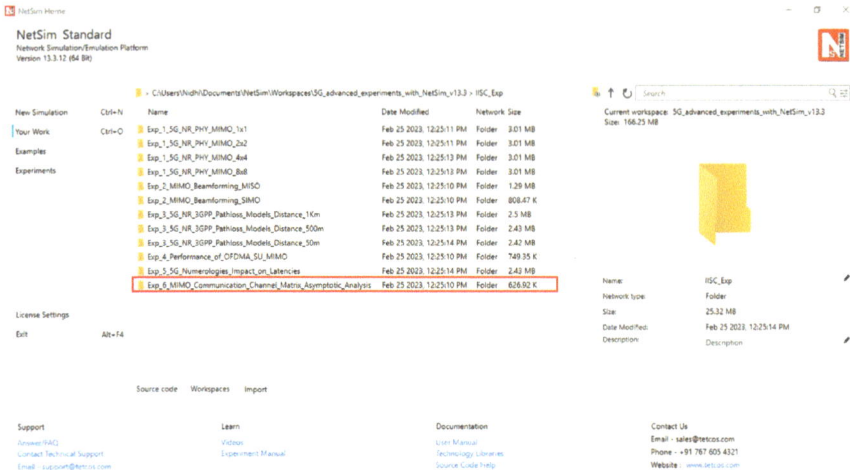

Fig. 12.4 Your Work window

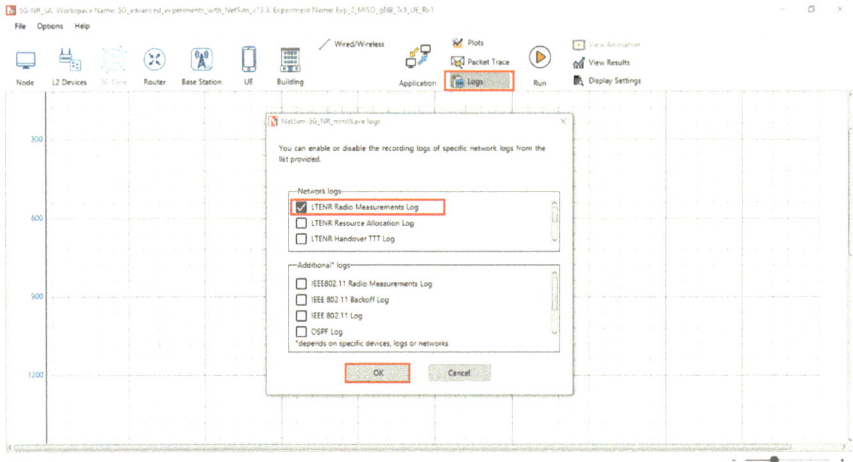

Fig. 12.5 Enabling LTENR radio measurement log file

Steps to Log the Results for Part I: Asymptotic Mean Condition Number Evaluation:

Steps to calculate the condition number through Eigenvalue (beamforming gain linear):

1. In the results window shown in Fig. 12.6, expand the Log Files option in the left panel and select LTENR_Radio_Measurements_Log.csv file.
2. This will open a.csv file, which logs the beamforming gain over time, as shown in Fig. 12.7.

Fig. 12.6 NetSim Results window showing access to the log files

PathLoss(dB)	ShadowFadingLoss(dB)	O2I_Loss(dBm)	Additional_Loss(dB)	Rx_Power(dBm)	SNR(dB)	SINR(dB)	InterferencePower(dBm)	BeamFormingGain(dB)	CQI Index	MCS Index
92.383716	-1.750078	0	0	-26.551238	77.276719	N/A	N/A	24.0824	N/A	N/A
92.383716	-1.750078	0	0	-26.551238	77.276719	N/A	N/A	24.0824	N/A	N/A
92.383716	3.621575	0	0	-96.525748	7.302208	7.302208	-1000	-28.479258	9	20
92.383716	3.621575	0	0	-72.22222	31.605736	31.605736	-1000	-4.17573	15	28
92.383716	3.621575	0	0	-68.292218	35.535738	35.535738	-1000	-0.245728	15	28
92.383716	3.621575	0	0	-65.819065	38.008891	38.008891	-1000	2.227425	15	28
92.383716	3.621575	0	0	-63.611513	40.216443	40.216443	-1000	4.434977	15	28
92.383716	3.621575	0	0	-60.831517	42.996439	42.996439	-1000	7.214973	15	28
92.383716	3.621575	0	0	-59.823436	44.00452	44.00452	-1000	8.223054	15	28
92.383716	3.621575	0	0	-57.892893	45.935064	45.935064	-1000	10.153598	15	28
92.383716	3.621575	0	0	-56.457741	47.370215	47.370215	-1000	11.588749	15	28
92.383716	3.621575	0	0	-56.094415	47.733541	47.733541	-1000	11.952075	15	28
92.383715	3.621575	0	0	-56.107734	47.720222	47.720222	-1000	11.938756	15	28
92.383716	3.621575	0	0	-55.085545	48.742411	48.742411	-1000	12.960945	15	28
92.383716	3.621575	0	0	-53.303162	50.524794	50.524794	-1000	14.743328	15	28
92.383716	3.621575	0	0	-52.170278	51.657678	51.657678	-1000	15.876212	15	28
92.383716	3.621575	0	0	-50.977606	52.85035	52.85035	-1000	17.068884	15	28

Fig. 12.7 LTENR radio measurements log file created after simulation

3. To change the beamforming gain from dB scale to linear, we need to use the formula:

$$\text{Eigen Value(Beam Forming Gain, Linear)} = 10^{\left(\frac{\text{Beam Forming Gain}_{dB}}{10}\right)},$$

and to do this, in a new column, enter the following function to calculate the linear beamforming gain:

$$= \text{POWER}\big(10, \big[@\big[\text{Beam Forming Gain}(dB)\big]\big]/10\big).$$

4. Now go to Insert, select Pivot table option, and then select new sheet option, and click OK, as shown in Fig. 12.8.
5. Drag and drop the channel and LAYER_ID field to filter block. Filter the channel to only PDSCH since we have considered a DL application from server to UE. Similarly, drag and drop the linear beamforming gain to the Values field and time to the Rows field, as shown in Fig. 12.9.

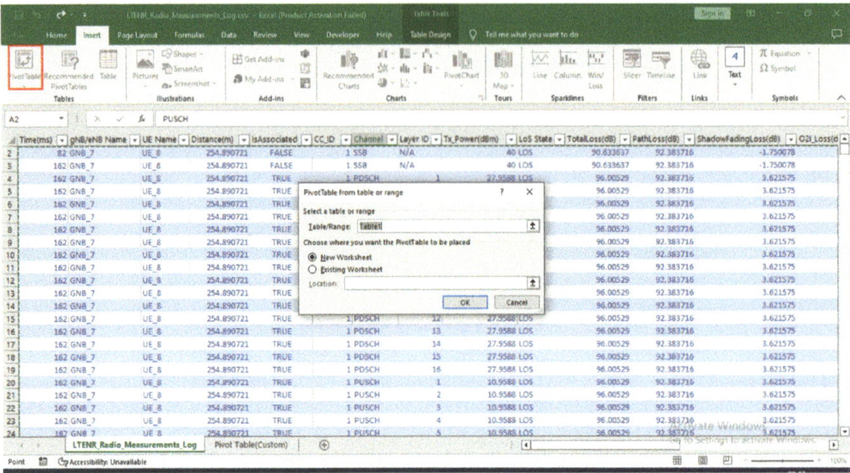

Fig. 12.8 Creating a pivot table

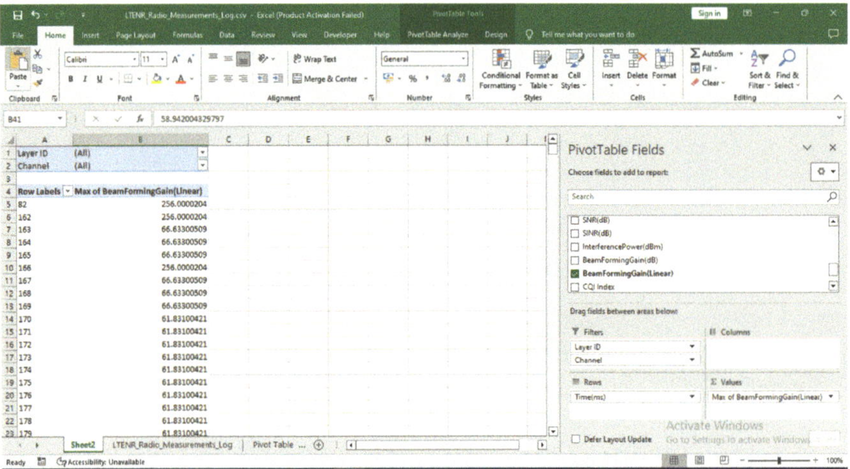

Fig. 12.9 Pivot table for LTENR radio measurements Log file showing the filtering process of the DL/UL

6. Now, filter the Layer_Id to layer 1 as shown in Fig. 12.10. Copy the beamforming values and paste the values in the new sheet with the column name as Layer 1.
7. Similarly, filter the LAYER_ID as 16 and copy all the eigenvalues and paste them in the new sheet with column name Layer 16 as shown in Fig. 12.11.
8. In the next column, enter the formula $= [@\text{Layer16}]/[@\text{Layer1}]$ to compute EV_max/EV_min. Note that the eigenvalues of Layer 16 are λ_{max} while the eigenvalue of Layer 1 is λ_{min}, and rename the column containing this ratio suitably, as shown in Fig. 12.12.

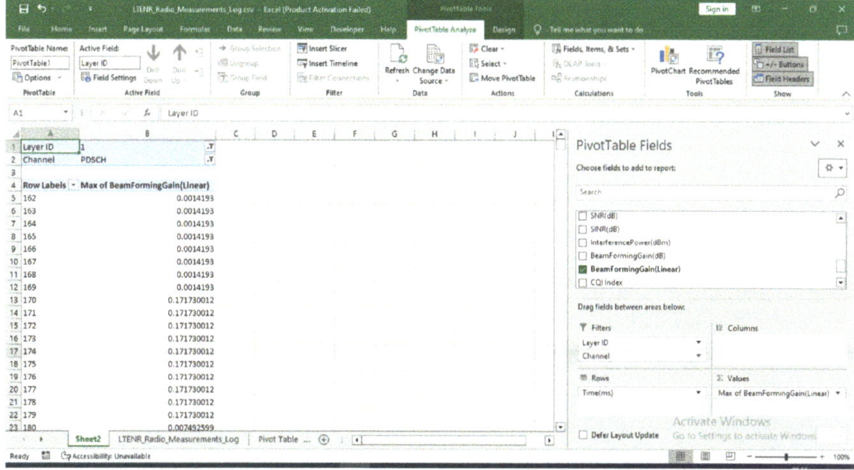

Layer ID	1
Channel	PDSCH

Row Labels	Max of BeamFormingGain(Linear)
162	0.0014193
163	0.0014193
164	0.0014193
165	0.0014193
166	0.0014193
167	0.0014193
168	0.0014193
169	0.0014193
170	0.171730012
171	0.171730012
172	0.171730012
173	0.171730012
174	0.171730012
175	0.171730012
176	0.171730012
177	0.171730012
178	0.171730012
179	0.171730012
180	0.007492599

Fig. 12.10 LTENR radio measurements Log file showing eigenvalue obtained for layer 1

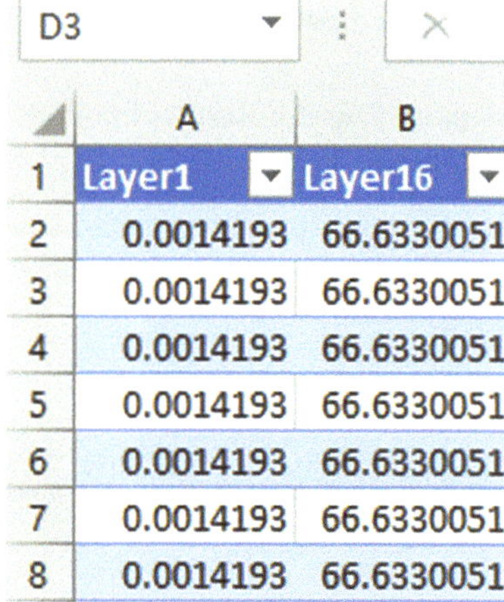

Layer1	Layer16
0.0014193	66.6330051
0.0014193	66.6330051
0.0014193	66.6330051
0.0014193	66.6330051
0.0014193	66.6330051
0.0014193	66.6330051
0.0014193	66.6330051

Fig. 12.11 Layer 1 and layer 16 eigenvalues in the new table

C13 =[@Layer16]/[@Layer1]

Layer1	Layer16	EV_Max_EV_Min
0.0014193	66.63300509	46947.79505
0.0014193	66.63300509	46947.79505
0.0014193	66.63300509	46947.79505
0.0014193	66.63300509	46947.79505
0.0014193	66.63300509	46947.79505

Layer1	Layer16	EV_Max_EV_Min
0.0014193	66.6330051	46947.79505
0.0014193	66.6330051	46947.79505
0.0014193	66.6330051	46947.79505
0.0014193	66.6330051	46947.79505
0.0014193	66.6330051	46947.79505
0.0014193	66.6330051	46947.79505

Fig. 12.12 Obtaining the ratio $\frac{\lambda_{max}}{\lambda_{min}}$

D13 =SQRT([@[EV_Max_EV_Min]])

	A	B	C	D
1	Layer1	Layer16	EV_Max_EV_Min	Condition_Number
2	0.0014193	66.63300509	46947.79505	216.6743987
3	0.0014193	66.63300509	46947.79505	216.6743987
4	0.0014193	66.63300509	46947.79505	216.6743987
5	0.0014193	66.63300509	46947.79505	216.6743987
6	0.0014193	66.63300509	46947.79505	216.6743987
7	0.0014193	66.63300509	46947.79505	216.6743987

Layer1	Layer16	EV_Max_EV_Min	Condition_Number
0.0014193	66.63300509	46947.79505	216.6743987
0.0014193	66.63300509	46947.79505	216.6743987
0.0014193	66.63300509	46947.79505	216.6743987
0.0014193	66.63300509	46947.79505	216.6743987
0.0014193	66.63300509	46947.79505	216.6743987
0.0014193	66.63300509	46947.79505	216.6743987
0.0014193	66.63300509	46947.79505	216.6743987

Fig. 12.13 Obtaining the condition numbers

9. In the next column, enter the formula = SQRT([@[EV_Max_EV_Min]]). This will calculate square root of the ratio $\sqrt{\frac{\lambda_{\max}}{\lambda_{\min}}}$, which computes the required condition number, as shown in Fig. 12.13.
10. In a new cell, enter the formula = AVERAGE(Table2[Condition_Number]) to calculate the average of the condition number across several channel realizations, as shown in Fig. 12.14.
11. Similarly, enter the formula, = VAR.*P*(Table2[Condition_Number]) in a new cell, to calculate the variance of condition number, as shown in Fig. 12.15.
12. Repeat steps 1 to 11 with varying Tx antenna count in gNB as 32, 64, and 128. Note down the mean and variance of the condition number.

Steps to Log the Results for Part II: Distribution of the Asymptotic Condition Number.

Steps to plot the histogram of condition number:

1. Calculate the condition number using the LTENR_Radio_Measurements_Log, as explained above.

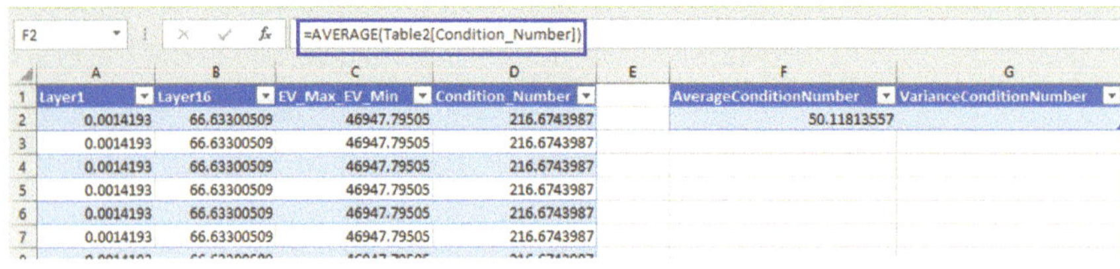

F2 =AVERAGE(Table2[Condition_Number])

	A	B	C	D	E	F	G
1	Layer1	Layer16	EV_Max_EV_Min	Condition_Number		AverageConditionNumber	VarianceConditionNumber
2	0.0014193	66.63300509	46947.79505	216.6743987		50.11813557	
3	0.0014193	66.63300509	46947.79505	216.6743987			
4	0.0014193	66.63300509	46947.79505	216.6743987			
5	0.0014193	66.63300509	46947.79505	216.6743987			
6	0.0014193	66.63300509	46947.79505	216.6743987			
7	0.0014193	66.63300509	46947.79505	216.6743987			

Fig. 12.14 Obtaining the average of the condition numbers

G2 =VAR.P(Table2[Condition_Number])

	A	B	C	D	E	F	G
1	Layer1	Layer16	EV_Max_EV_Min	Condition_Number		AverageConditionNumber	VarianceConditionNumber
2	0.0014193	66.63300509	46947.79505	216.6743987		50.11813557	3168.078574
3	0.0014193	66.63300509	46947.79505	216.6743987			
4	0.0014193	66.63300509	46947.79505	216.6743987			
5	0.0014193	66.63300509	46947.79505	216.6743987			
6	0.0014193	66.63300509	46947.79505	216.6743987			
7	0.0014193	66.63300509	46947.79505	216.6743987			

Fig. 12.15 Obtaining the variance of the condition numbers

2. In a new column, divide the Condition_Number by 16 (or in general by $n = \min(N_r, N_t)$) using the following excel function:

$$= [@Condition\_Number]/16)$$

3. Click on Insert- > Pivot Table, drag and drop Column 1 to the Rows field.
4. Copy the values in the Row Labels column.
5. In MATLAB, create a new file, create an array **Condition_Number_array**, and paste the values to it as.

Condition_Number_array = [c1.
c2.
c3.
....
cn];

6. Now use below MATLB code to plot the normalized histogram plot, along with the function given in Sect. 12.4.2.

Program 1: MATLAB code for plotting the condition number from NetSim and comparing against the asymptotic PDF from theoretical analysis

```
hold on;
c = histogram(Condition_Number_array,'Normalization','probability');
x = 0:0.1:50; %x varies from 0 to 50 in steps of 0.2
y = (8./x.^3).*(exp(-4./x.^2));
plot(x,y,'r');
hold off;
% For a CDF plot, use following MATLAB function:
cdfplot(Condition_Number_array);
```

Steps to Log the Results for Part III: Marchenko-Pastur Distribution

Steps to plot the histogram:

1. Create a scenario with Tx antenna count as 128 and Rx antenna count as 16.
2. Now open the file LTENR_Radio_Measurements_Log.csv file.
3. Filter the PDSCH/PUSCH to PDSCH.
4. Now, in LTENR Radio Measurements Log file, compute the eigenvalues (linear beamforming gain) for all layers, Layer Id 1 to Layer Id 16, as explained previously.
5. Select the eigenvalues in the beamforming gain column and copy them to the clipboard.
6. Create a new file in MATLAB. Create an array, Eigen_value_array, and paste the copied values from the 5G parameter log csv file as shown.

Eigen_value_array = [ev1
ev2
...

evN];

1.7 Use the MATLAB code below to plot the MP distribution function along with the normalized histogram.

For the MP distribution function, the x varies from 0.418 to 1.832 in steps of 0.001.

Program 2 MATLAB code for plotting the MP distribution with $y = \frac{1}{8}$ and the histogram of the eigenvalues from NetSim simulation results

```
x = 0.418:0.001:1.832;
y = (1.27324./x).*sqrt(0.5 - (x-9/8).^2);
hold on;
plot(x,y);
histogram(Eigen_value_array /128,'Normalization','pdf');
hold off;
```

Reference

Edelman A (1988) Eigenvalues and condition numbers of random matrices. SIAM J Matrix Anal Appl 9(4):543–560

Zeitfracht Medien GmbH
Ferdinand-Jühlke-Straße 7
99095 Erfurt, Deutschland
produktsicherheit@kolibri360.de